Introduction to Nonlinear Thermomechanics

Theory and Finite-Element Solutions

Andrzej Służalec

Introduction to Nonlinear Thermomechanics

Theory and Finite-Element Solutions

With 78 Figures

Springer-Verlag
London Berlin Heidelberg New York
Paris Tokyo Hong Kong
Barcelona Budapest

Andrzej Służalec
Technical University of Częstochowa,
Częstochowa, Poland

ISBN 3-540-19703-6 Springer-Verlag Berlin Heidelberg New York
ISBN 0-387-19703-6 Springer-Verlag New York Berlin Heidelberg

British Library Cataloguing in Publication Data
Służalec, Andrzej
  Introduction to nonlinear thermomechanics.
  I. Title
  536.7
ISBN 3540197036

Library of Congress Cataloging-in-Publication Data
Służalec, Andrzej 1955–
  Introduction to nonlinear thermomechanics/Andrzej Służalec.
    p.   cm.
  Includes index.
  ISBN 0-387-19703-6 (U.S.)
  1. Thermodynamics. 2. Nonlinear mechanics. I. Title.
QC311.2.S96 1991
620.1′121—dc20

91-16574
CIP

Apart from any fair dealing for the purposes of research or private study, or criticism or review, as permitted under the Copyright, Designs and Patents Act 1988, this publication may only be reproduced, stored or transmitted, in any form or by any means, with the prior permission in writing of the publishers, or in the case of reprographic reproduction in accordance with the terms of licences issued by the Copyright Licensing Agency. Enquiries concerning reproduction outside those terms should be sent to the publishers.

© Springer-Verlag London Limited 1992
Printed in Germany

The publisher makes no representation, express or implied, with regard to the accuracy of the information contained in this book and cannot accept any legal responsibility or liability for any errors or omissions that may be made.

Typeset by Macmillan India Ltd., Bangalore 560 025
69/3830-543210 Printed on acid-free paper

PREFACE

The interrelation of temperature and deformation appears in mechanics in various forms. A thermal field influences the material properties, modifies the extent of plastic zones, results in ratchetting during cyclic heating etc.; conversely, deformation induces changes in temperature distribution. From the stress analysis point of view the thermal effects can be studied at two levels, depending on whether uncoupled or coupled theories of thermomechanical response have to be employed. A majority of technologically important problems can be satisfactorily studied within an uncoupled theory. In such an approach the temperature enters the stress–strain relation through the thermal dilatation and possibly influences the material constants. The heat conduction equation and the relations governing the stress field are considered separately. There exist, however, many instances when coupling of thermal and deformation states should be considered because of its engineering importance; for example in nuclear engineering and metal-forming industry.

The effective solutions of complex thermomechanical problems only recently became possible. In the last two decades one observes the vigorous development of effective nonlinear methods of structural analysis. Many industries such as the automobile, aircraft, nuclear, aerospace or ship industry are experiencing a rapidly-growing need for the analytical tools to handle complex problems of modern technology. Efficient analytical methods for combining geometrically, materially and thermally nonlinear structural problems are needed because experimental testing in such cases is often prohibitively expensive or physically impossible. Recent advances in computer technology make it possible to perform extensive calculations with great accuracy, at significantly reduced execution times and at reasonable cost.

The intention of this book is to reveal and discuss some aspects of nonlinear thermomechanics, with special attention focused upon numerical implementations for the problems undertaken. Thus, after the theoretical concepts, possible finite-element formulations and numerical solution strategy are discussed. Many examples are used to illustrate application to engineering problems. There are relatively few published review articles available on solving complex problems of nonlinear thermomechanics. There exist some monographs devoted entirely to the finite-element analysis of materially nonlinear solid mechanics problems (Oden 1972; Bathe 1982), but nonlinear thermomechanical problems are treated rather incidentally in these books.

This book has been divided into six parts. It contains basic considerations and notions in thermomechanics (Part I); recalls fundamentals of elasticity and plasticity theory (Part II); and analyses small strain thermo-elasto-plasticity (Part III), creep (Part IV), finite strains (Part V) and coupled thermomechanical processes (Part VI). The book is written using a simple mathematical language requiring only an elementary background in mathematical and tensor calculus in the hope that this will render it available to a wider readership.

The author is conscious that, in illustrating to the reader the solution of problems in nonlinear thermomechanics, his choice of topics is by no means comprehensive, but rather reflects his own experience in this field.

Częstochowa Andrzej Służalec
April 1991

CONTENTS

PART I. BASIC CONSIDERATIONS AND NOTIONS

1 A State of Stress and Strain 3
   1.1 Stress 3
   1.2 Strain 8

2 Finite Strains 12
   2.1 Finite Strain Tensor in Material and Spatial
        Descriptions 12
   2.2 Deformation Rate Tensor 19
   2.3 Stress Measures 24
   2.4 Final Remarks 26

3 Temperature 27
   3.1 Heat Conduction 27
   3.2 Heat Convection 28
   3.3 Heat Radiation 29
   3.4 Temperature Field in a Heat-Conducting Body 30
   3.5 Navier–Stokes Equation 33

4 Thermodynamical Considerations 35
   4.1 Thermomechanical Process 35
   4.2 Formulation of the Constitutive Law 36

PART II. FUNDAMENTALS OF ELASTICITY AND PLASTICITY THEORY

5 Stress–Strain Curve 45

6 Elasticity 48

7 Plasticity 50

- 7.1 Idealization of Tension Test 50
- 7.2 Ideal Plasticity Theories 52
  - 7.2.1 Yield Criteria 52
  - 7.2.2 Hencky–Iljuszyn Deformation Theory 53
  - 7.2.3 Plastic Flow Theory 55
  - 7.2.4 Comparison of Flow Theory and Deformation Theory 56
  - 7.2.5 Ideal Plasticity Theory for Finite Deformations .. 58

8 Work-Hardening Equation 60

- 8.1 Drucker Postulate 60
  - 8.1.1 Stability of Plastic Material in the Drucker Sense. 60
  - 8.1.2 Associated Plastic Flow 61
- 8.2 Yield Surfaces for Work-Hardening Materials 63
  - 8.2.1 Experimental Results 63
  - 8.2.2 Isotropic Hardening 65
  - 8.2.3 Kinematic Hardening 66

PART III. SMALL STRAIN THERMO-ELASTO-PLASTICITY

9 Equations for Thermo-Elasto-Plasticity 73

- 9.1 Isotropic Hardening 73
- 9.2 Kinematic Hardening 76
- 9.3 Elasto-Visco-Plasticity 78

10 Finite-Element Solution 80

- 10.1 Finite-Element Solution of Heat Flow Equations 80
  - 10.1.1 Weighted Residual Method 80
  - 10.1.2 Variational Formulation 82
  - 10.1.3 Time Integration Schemes for Nonlinear Heat Conduction 84
  - 10.1.4 Stability Analysis 88
- 10.2 Finite-Element Solution of Navier–Stokes Equations 90
- 10.3 Modelling of the Phase Change Process 92
- 10.4 Examples of Thermal Problems 94
  - 10.4.1 Heat Flow with Phase Change 94
  - 10.4.2 Navier–Stokes Equations 97
- 10.5 Finite-Element Solution of Thermo-Elasto-Plastic Problems 100
  - 10.5.1 Variational Formulation 100
  - 10.5.2 Integration 104

       10.5.3 Methods of Iterative Accumulation 105
       10.5.4 Tangent Stiffness Matrices 106
  10.6 Examples of Thermo-Elasto-Plastic Analyses 107

PART IV. CREEP

11 Theoretical Background to Creep 117

  11.1 Creep and Relaxation Tests 117
  11.2 Creep at Constant Uniaxial Stress 117
       11.2.1 Time Functions 117
       11.2.2 Stress Functions 119
       11.2.3 Temperature Functions 119
       11.2.4 Stress and Time Functions 120
  11.3 Creep Theories with Time-Dependent Uniaxial
      Stress 120
       11.3.1 Total Strain Theory 120
       11.3.2 Time Hardening Theory 121
       11.3.3 Strain Hardening Theory 121
       11.3.4 Heredity Theory 123
  11.4 Creep Theories in Complex Stress State 124
       11.4.1 Creep Theory of Deformational Type 124
       11.4.2 Flow Theories and Creep Potential 124
       11.4.3 Generalization of Strain Hardening Theory . . . 126

12 Creep Rupture 128

  12.1 Experimental Studies 128
  12.2 Ductile Rupture Theories 129
  12.3 Brittle Rupture Theories 131
  12.4 Rupture of Mixed Type 132

**13 Constitutive Equations for Thermo-Elasto-Plastic and
    Creep Analysis** 134

14 Finite-Element Formulation 135

  14.1 Matrix Equation for Thermo-Elasto-Plastic and
      Creep Problems 135
  14.2 Remarks on Solution Procedures 136
  14.3 Examples 139

PART V. FINITE STRAINS

15 Finite Strain Models 151

16 Constitutive Equations 152

16.1 Non-Isothermal Plastic Flow 152
16.2 Multiplicative Decomposition of the Deformation Gradient 155

17 Finite-Element Formulation for Non-Isothermal Plastic Flow 157

17.1 Total Lagrange Formulation 157
17.2 Updated Lagrange and Updated Lagrange–Jaumann Formulations 158
17.3 Updated Lagrange–Hughes Formulation 160

PART VI. COUPLED THERMO-PLASTICITY

18 Equations of Coupled Thermo-Plasticity 163

18.1 Heat Transfer Equations....................... 163
18.2 Finite-Element Formulation for the Heat Flow Equation 165
18.3 Internal Dissipation Function................... 166
18.4 Stress–Strain Relations in Coupled Thermo-Plasticity...................................... 167
    18.4.1 Thermo-Elasto-Plastic Model Based on Additive Decomposition of Strain 167
    18.4.2 Thermo-Rigid Plastic and Thermo-Rigid Visco-Plastic Models 168
    18.4.3 Remarks on Other Models................. 169
18.5 Coupled Thermomechanical Algorithm 169
18.6 Examples 170

References and Further Reading 177

Subject Index 185

PART I
BASIC CONSIDERATIONS AND NOTIONS

1 A State of Stress and Strain

1.1 Stress

In the course of deformation of real solids, on account of volume and geometrical shape changes, interactions between molecules come into being that oppose these changes. Considering inside a body a surface element ΔS with normal $\boldsymbol{n}$ on which a resultant internal force $\Delta \boldsymbol{P}$ acts, we define the stress vector at a point lying on this surface as

$$p^{(n)} = \lim_{\Delta S \to 0} \frac{\Delta \boldsymbol{P}}{\Delta S} = \frac{d\boldsymbol{P}}{dS}. \tag{1.1}$$

This vector can be resolved into a component in the direction of the normal σ and a component lying in the plane of the element surface τ. By a state of stress at a point is meant a set of stresses acting on all elements of the surface dS that contain this point. Introducing the Cartesian coordinate system x, y, z and selecting, at the point considered, elements of surfaces $dS^{(i)}$ in such a way that their normals are parallel to the system axis (Fig. 1.1) we get three stresses $\boldsymbol{p}^{(i)}$ ($i = x, y, z$), defining the state of stress at that point

$$\boldsymbol{p}^{(n)} = \boldsymbol{p}^{(x)} n_x + \boldsymbol{p}^{(y)} n_y + \boldsymbol{p}^{(z)} n_z = \boldsymbol{p}^{(k)} n_k. \tag{1.2}$$

Each of these vectors can be resolved into three components coincident with the directions of the system axes. Finally the state of stress at a point is described by nine components σ_{ij}, that form a second order tensor called the stress tensor. The vector equations (1.2) can be expressed in the form of three equations determining the vector components $\boldsymbol{p}^{(n)}$

$$\begin{aligned} p_x^{(n)} &= \sigma_{xx} n_x + \sigma_{yx} n_y + \sigma_{zx} n_z, \\ p_y^{(n)} &= \sigma_{xy} n_x + \sigma_{yy} n_y + \sigma_{zy} n_z, \\ p_z^{(n)} &= \sigma_{xz} n_x + \sigma_{yz} n_y + \sigma_{zz} n_z, \\ n_x &= \cos(n, x), \quad n_y = \cos(n, y), \quad n_z = \cos(n, z) \end{aligned} \tag{1.3}$$

or, using summation convention

$$p_i^{(n)} = \sigma_{ji} n_j. \tag{1.4}$$

The majority of authors use two basic methods for stress tensor component notation: a two-index mathematical notation, making possible the use of the Einstein summation convention; and engineering notation in which normal

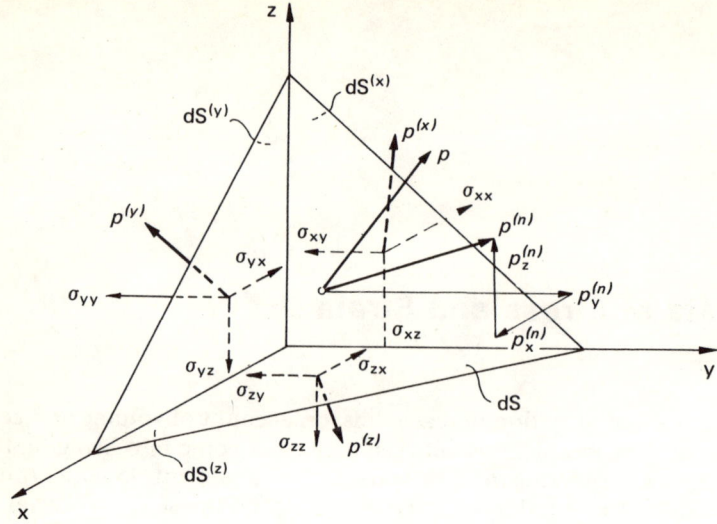

Figure 1.1. Stress vector and stress tensor components in mathematical notation.

components are denoted by the symbol σ with one index, and tangent components by the symbol τ with two indices (Fig. 1.2)

$$\sigma_{ij} = \begin{Bmatrix} \sigma_{11} & \sigma_{21} & \sigma_{31} \\ \sigma_{12} & \sigma_{22} & \sigma_{32} \\ \sigma_{13} & \sigma_{23} & \sigma_{33} \end{Bmatrix} = \begin{Bmatrix} \sigma_{xx} & \sigma_{yx} & \sigma_{zx} \\ \sigma_{xy} & \sigma_{yy} & \sigma_{zy} \\ \sigma_{xz} & \sigma_{yz} & \sigma_{zz} \end{Bmatrix} = \begin{Bmatrix} \sigma_{x} & \sigma_{yx} & \sigma_{zx} \\ \sigma_{xy} & \sigma_{y} & \sigma_{zy} \\ \sigma_{xz} & \sigma_{yz} & \sigma_{z} \end{Bmatrix} \qquad (1.5)$$

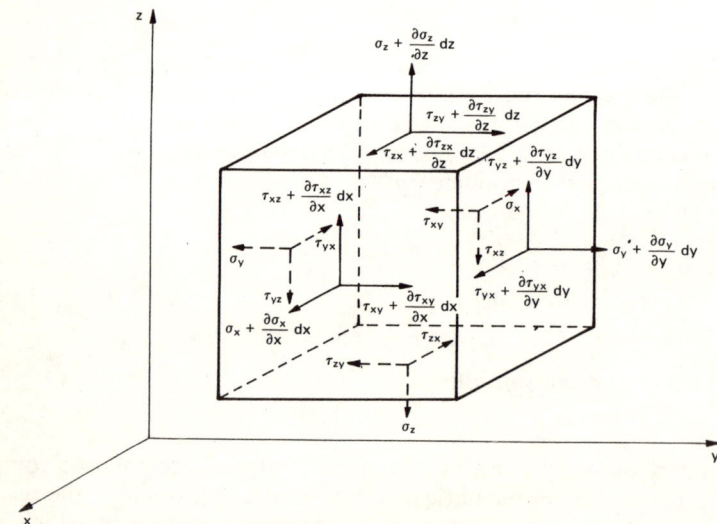

Figure 1.2. Stress tensor components in engineering notation; internal equilibrium equations.

The principal stresses σ_i are eigenvalues of σ_{ij}. They can be calculated as roots of the equation

$$\sigma^3 - J_{1\sigma}\sigma^2 + J_{2\sigma}\sigma - J_{3\sigma} = 0 \tag{1.6}$$

where $J_{1\sigma}, J_{2\sigma}, J_{3\sigma}$ are homogeneous functions of the stress components σ_{ij} that are invariant with respect to coordinate system transformation. They are called principal invariants of the stress tensor.

$$J_{1\sigma} = \sigma_{ii} = \sigma_x + \sigma_y + \sigma_z,$$

$$J_{2\sigma} = \tfrac{1}{2}(\sigma_{ii}\sigma_{jj} - \sigma_{ij}\sigma_{ji}) = \begin{vmatrix} \sigma_x & \tau_{yx} \\ \tau_{xy} & \sigma_y \end{vmatrix} + \begin{vmatrix} \sigma_y & \tau_{zy} \\ \tau_{yz} & \sigma_z \end{vmatrix} + \begin{vmatrix} \sigma_z & \tau_{xz} \\ \tau_{zx} & \sigma_x \end{vmatrix}$$

$$= \sigma_x\sigma_y + \sigma_y\sigma_z + \sigma_z\sigma_x - (\tau_{xy}^2 + \tau_{yz}^2 + \tau_{zx}^2), \tag{1.7}$$

$$J_{3\sigma} = \tfrac{1}{6}e_{ijk}e_{pqr}\sigma_{ip}\sigma_{jq}\sigma_{kr} = \begin{vmatrix} \sigma_x & \tau_{yx} & \tau_{zx} \\ \tau_{xy} & \sigma_y & \tau_{zy} \\ \tau_{xz} & \tau_{yz} & \sigma_z \end{vmatrix}$$

$$= \sigma_x\sigma_y\sigma_z + 2\tau_{xy}\tau_{yz}\tau_{zx} - (\sigma_x\tau_{yz}^2 + \sigma_y\tau_{zx}^2 + \sigma_z\tau_{xy}^2).$$

Permutation symbol e_{ijk} takes the values

$$e_{ijk} = \begin{cases} 0 - \text{if any two indices are identical,} \\ +1 - \text{if all indices are different and permutation is even,} \\ -1 - \text{if all indices are different and permutation is odd.} \end{cases} \tag{1.8}$$

This means

$$\begin{aligned} e_{111} &= e_{112} = e_{122} = \ldots = 0, \\ e_{123} &= e_{231} = e_{312} = +1, \\ e_{213} &= e_{132} = e_{321} = -1. \end{aligned} \tag{1.9}$$

If $\sigma_1, \sigma_2, \sigma_3$ are principal stresses, then Eq. (1.6) can be written in the form

$$(\sigma - \sigma_1)(\sigma - \sigma_2)(\sigma - \sigma_3) = 0, \tag{1.10}$$

and principal invariants can be expressed by the simple formulas

$$\begin{aligned} J_{1\sigma} &= \sigma_1 + \sigma_2 + \sigma_3, \\ J_{2\sigma} &= \sigma_1\sigma_2 + \sigma_2\sigma_3 + \sigma_3\sigma_1, \\ J_{3\sigma} &= \sigma_1\sigma_2\sigma_3. \end{aligned} \tag{1.11}$$

A stress tensor can be written as the sum of the spherical $\sigma_m \delta_{ij}$ and the deviatoric s_{ij} stresses according to the equation

$$\sigma_{ij} = s_{ij} + \sigma_m \delta_{ij}, \tag{1.12}$$

where the Kronecker symbol δ_{ij} is defined as

$$\delta_{ij} = \begin{cases} 0 & (i \neq j), \\ 1 & (i = j), \end{cases} \tag{1.13}$$

and σ_m is the mean stress

$$\sigma_m = \tfrac{1}{3}\sigma_{kk} = \tfrac{1}{3}(\sigma_x + \sigma_y + \sigma_z) = \tfrac{1}{3}J_{1\sigma}. \tag{1.14}$$

Deviatoric stress has the form

$$s_{ij} = \begin{Bmatrix} \sigma_x - \sigma_m & \tau_{yx} & \tau_{zx} \\ \tau_{xy} & \sigma_y - \sigma_m & \tau_{zy} \\ \tau_{xz} & \tau_{yz} & \sigma_z - \sigma_m \end{Bmatrix}$$

$$= \begin{Bmatrix} \frac{2}{3}\sigma_x - \frac{1}{3}(\sigma_y + \sigma_z) & \tau_{yx} & \tau_{zx} \\ \tau_{xy} & \frac{2}{3}\sigma_y - \frac{1}{3}(\sigma_z + \sigma_x) & \tau_{zy} \\ \tau_{xz} & \tau_{yz} & \frac{2}{3}\sigma_z - \frac{1}{3}(\sigma_x + \sigma_y) \end{Bmatrix}. \quad (1.15)$$

Principal invariants of the deviatoric stress we obtain by replacing σ by $(s + \frac{1}{3}J_{1\sigma})$ in Eq. (1.6). Then the equation takes the form

$$s^3 - J_{1s}s^2 + J_{2s}s - J_{3s} = 0 \quad (1.16)$$

where J_{1s}, J_{2s}, J_{3s} are principal invariants of the deviatoric stress. They are determined by the following scalar functions of the deviatoric stress components s_{ij}

$$\begin{aligned} J_{1s} &= s_{ii} = s_{xx} + s_{yy} + s_{zz} = 0, \\ J_{2s} &= -\tfrac{1}{2}s_{ij}s_{ji} = -\tfrac{1}{2}[s_{xx}^2 + s_{yy}^2 + s_{zz}^2 + 2(s_{xy}^2 + s_{yz}^2 + s_{zx}^2)], \\ J_{3s} &= \tfrac{1}{3}s_{ij}s_{jk}s_{ki} = \tfrac{1}{3}[s_{xx}^3 + s_{yy}^3 + s_{zz}^3 + 3(s_{xx}s_{xy}^2 + s_{xx}s_{xz}^2 \\ &\quad + s_{yy}s_{yz}^2 + s_{yy}s_{yx}^2 + s_{zz}s_{zx}^2 + s_{zz}s_{zy}^2) + 6s_{xy}s_{yz}s_{zx}], \end{aligned} \quad (1.17)$$

or by principal values of the deviatoric stress s_i

$$\begin{aligned} J_{2s} &= -\tfrac{1}{2}(s_1^2 + s_2^2 + s_3^2) = s_1s_2 + s_2s_3 + s_3s_1, \\ J_{3s} &= \tfrac{1}{3}(s_1^3 + s_2^3 + s_3^3) = s_1s_2s_3. \end{aligned} \quad (1.18)$$

One can also express principal invariants of the deviatoric stress as a function of stress tensor components σ_{ij} or its eigenvalues σ_i

$$\begin{aligned} J_{2s} &= -\tfrac{1}{3}[\sigma_x^2 + \sigma_y^2 + \sigma_z^2 - (\sigma_x\sigma_y + \sigma_y\sigma_z + \sigma_z\sigma_x) + 3(\tau_{xy}^2 + \tau_{yz}^2 + \tau_{zx}^2)] \\ &= -\tfrac{1}{6}[(\sigma_x - \sigma_y)^2 + (\sigma_y - \sigma_z)^2 + (\sigma_z - \sigma_x)^2 + 6(\tau_{xy}^2 + \tau_{yz}^2 + \tau_{zx}^2)] \\ &= -\tfrac{1}{3}[\sigma_1^2 + \sigma_2^2 + \sigma_3^2 - (\sigma_1\sigma_2 + \sigma_2\sigma_3 + \sigma_3\sigma_1)], \\ J_{3s} &= \tfrac{4}{9}\sigma_x\sigma_y\sigma_z - \tfrac{1}{9}(\sigma_x^2\sigma_y + \sigma_x^2\sigma_z + \sigma_y^2\sigma_x + \sigma_y^2\sigma_z + \sigma_z^2\sigma_x + \sigma_z^2\sigma_y) \\ &\quad + \tfrac{2}{27}(\sigma_x^3 + \sigma_y^3 + \sigma_z^3) - \tfrac{2}{3}(\sigma_x\tau_{yz}^2 + \sigma_y\tau_{zx}^2 + \sigma_z\tau_{xy}^2) \\ &\quad + \tfrac{1}{3}(\sigma_x\tau_{xy}^2 + \sigma_x\tau_{xz}^2 + \sigma_y\tau_{yx}^2 + \sigma_y\tau_{yz}^2 + \sigma_z\tau_{zx}^2 + \sigma_z\tau_{zy}^2) + 2\tau_{xy}\tau_{yz}\tau_{zx} \\ &= \tfrac{4}{9}\sigma_1\sigma_2\sigma_3 - \tfrac{1}{9}(\sigma_1^2\sigma_2 + \sigma_1^2\sigma_3 + \sigma_2^2\sigma_1 + \sigma_2^2\sigma_3 + \sigma_3^2\sigma_1 + \sigma_3^2\sigma_2) \\ &\quad + \tfrac{2}{27}(\sigma_1^3 + \sigma_2^3 + \sigma_3^3). \end{aligned} \quad (1.19)$$

The simple relationships between principal invariants of the deviatoric stress J_{2s}, J_{3s} and principal invariants of the stress tensor $J_{1\sigma}, J_{2\sigma}, J_{3\sigma}$ hold

$$\begin{aligned} J_{2s} &= -\tfrac{1}{3}(J_{1\sigma}^2 - 3J_{2\sigma}), \\ J_{3s} &= \tfrac{1}{27}(2J_{1\sigma}^3 - 9J_{1\sigma}J_{2\sigma} + 27J_{3\sigma}). \end{aligned} \quad (1.20)$$

In plasticity theory a somewhat different scalar function of deviatoric stress or tensor stress components can be useful. It is called the effective stress and is expressed as

$$\bar{\sigma} = \bar{s} = \sqrt{\tfrac{3}{2} s_{ij} s_{ij}} = \sqrt{-3J_{2s}}$$

$$= \frac{1}{\sqrt{2}} \sqrt{(\sigma_{xx} - \sigma_{yy})^2 + (\sigma_{yy} - \sigma_{zz})^2 + (\sigma_{zz} - \sigma_{xx})^2 + 6(\sigma_{xy}^2 + \sigma_{yz}^2 + \sigma_{zx}^2)}$$

$$= \sqrt{\sigma_x^2 + \sigma_y^2 + \sigma_z^2 - \sigma_x \sigma_y - \sigma_y \sigma_z - \sigma_z \sigma_x + 3(\tau_{xy}^2 + \tau_{yz}^2 + \tau_{zx}^2)}. \tag{1.21}$$

The components of the stress tensor σ_{ij} must fulfil a set of equations of internal equilibrium

$$\frac{\partial \sigma_{ij}}{\partial x_j} + F_i = 0, \tag{1.22}$$

and conditions of shearing stress equivalence

$$\sigma_{ij} = \sigma_{ji}. \tag{1.23}$$

In Cartesian coordinates the internal equilibrium conditions take the form

$$\frac{\partial \sigma_x}{\partial x} + \frac{\partial \tau_{yx}}{\partial y} + \frac{\partial \tau_{zx}}{\partial z} + X = 0,$$

$$\frac{\partial \tau_{xy}}{\partial x} + \frac{\partial \sigma_y}{\partial y} + \frac{\partial \tau_{zy}}{\partial z} + Y = 0, \tag{1.24}$$

$$\frac{\partial \tau_{xz}}{\partial x} + \frac{\partial \tau_{yz}}{\partial y} + \frac{\partial \sigma_z}{\partial z} + Z = 0.$$

In the cylindrical coordinate system r, θ, z the equilibrium equation system is as follows

$$\frac{\partial \sigma_r}{\partial r} + \frac{1}{r}\frac{\partial \tau_{\theta r}}{\partial \theta} + \frac{\partial \tau_{zr}}{\partial z} + \frac{\sigma_r - \sigma_\theta}{r} + R = 0,$$

$$\frac{\partial \tau_{r\theta}}{\partial r} + \frac{1}{r}\frac{\partial \sigma_\theta}{\partial \theta} + \frac{\partial \tau_{z\theta}}{\partial z} + 2\frac{\tau_{r\theta}}{r} + \Theta = 0, \tag{1.25}$$

$$\frac{\partial \tau_{rz}}{\partial r} + \frac{1}{r}\frac{\partial \tau_{\theta z}}{\partial \theta} + \frac{\partial \sigma_z}{\partial z} + \frac{\tau_{rz}}{r} + Z = 0.$$

and in spherical coordinates r, θ, φ

$$\frac{\partial \sigma_r}{\partial r} + \frac{1}{r}\frac{\partial \tau_{\theta r}}{\partial \theta} + \frac{1}{r \sin \theta}\frac{\partial \tau_{\varphi r}}{\partial \varphi} + \frac{1}{r}[2\sigma_r - (\sigma_\theta + \sigma_\varphi) + \tau_{\theta r} \operatorname{ctg} \varphi] + R = 0,$$

$$\frac{\partial \tau_{r\theta}}{\partial r} + \frac{1}{r}\frac{\partial \sigma_\theta}{\partial \theta} + \frac{1}{r \sin \theta}\frac{\partial \tau_{\varphi \theta}}{\partial \varphi} + \frac{1}{r}[3\tau_{r\theta} + (\sigma_\theta - \sigma_\varphi)\operatorname{ctg} \theta] + \Theta = 0, \tag{1.26}$$

$$\frac{\partial \tau_{r\theta}}{\partial r} + \frac{1}{r}\frac{\partial \tau_{\theta \varphi}}{\partial \theta} + \frac{1}{r \sin \theta}\frac{\partial \sigma_\varphi}{\partial \varphi} + \frac{1}{r}[3\tau_{r\varphi} + 2\tau_{\varphi \theta} \operatorname{ctg} \theta] + \Phi = 0.$$

In the above equations $X, Y, Z; R, \Theta, Z$ and R, Θ, Φ denote the respective body force components in the directions of the coordinate system assumed.

1.2 Strain

The displacement of an arbitrary point of the body undergoing deformation can be expressed by the displacement vector $\boldsymbol{u}(u_1, u_2, u_3) = \boldsymbol{u}(u, v, w)$. In the analysis of small strains and with the additional assumption of small rotations, one introduces the second order symmetric tensor, called the infinitesimal strain tensor of deformation (Cauchy relations)

$$\varepsilon_{ij} = \frac{1}{2}\left(\frac{\partial u_i}{\partial x_j} + \frac{\partial u_j}{\partial x_i}\right). \tag{1.27}$$

As in the case of the stress tensor, two fundamental notations for the description of strain tensor components are used: two-index mathematical notation; and engineering notation, in which strain components are denoted by the symbol ε with respective index and/or by the symbol $\gamma/2$ with two indices corresponding to the plane in which the change of angle described by this deformation is carried out. So the infinitesimal strain tensor can be presented as

$$\varepsilon_{ij} = \begin{Bmatrix} \varepsilon_{11} & \varepsilon_{21} & \varepsilon_{31} \\ \varepsilon_{12} & \varepsilon_{22} & \varepsilon_{32} \\ \varepsilon_{13} & \varepsilon_{23} & \varepsilon_{33} \end{Bmatrix} = \begin{Bmatrix} \varepsilon_{xx} & \varepsilon_{yx} & \varepsilon_{zx} \\ \varepsilon_{xy} & \varepsilon_{yy} & \varepsilon_{zy} \\ \varepsilon_{xz} & \varepsilon_{yz} & \varepsilon_{zz} \end{Bmatrix} = \begin{Bmatrix} \varepsilon_x & \frac{1}{2}\gamma_{yx} & \frac{1}{2}\gamma_{zx} \\ \frac{1}{2}\gamma_{xy} & \varepsilon_y & \frac{1}{2}\gamma_{zy} \\ \frac{1}{2}\gamma_{xz} & \frac{1}{2}\gamma_{yz} & \varepsilon_z \end{Bmatrix}. \tag{1.28}$$

The Cauchy relations (1.27) using Cartesian coordinates are expressed as a set of six linear partial equations for u, v, w

$$\begin{aligned}
\varepsilon_x &= \frac{\partial u}{\partial x}, & \gamma_{xy} &= \frac{\partial u}{\partial y} + \frac{\partial v}{\partial x}, \\
\varepsilon_y &= \frac{\partial v}{\partial y}, & \gamma_{yz} &= \frac{\partial v}{\partial z} + \frac{\partial w}{\partial y}, \\
\varepsilon_z &= \frac{\partial w}{\partial z}, & \gamma_{zx} &= \frac{\partial w}{\partial x} + \frac{\partial u}{\partial z}.
\end{aligned} \tag{1.29}$$

In the cylindrical coordinate system r, θ, z, in which the displacement vector components are denoted by u_r, u_θ, u_z, the Cauchy relations take the form

$$\begin{aligned}
\varepsilon_r &= \frac{\partial u_r}{\partial r}, & \gamma_{r\theta} &= \frac{1}{r}\frac{\partial u_r}{\partial \theta} + \frac{\partial u_\theta}{\partial r} - \frac{u_\theta}{r}, \\
\varepsilon_\theta &= \frac{1}{r}\frac{\partial u_\theta}{\partial \theta} + \frac{u_r}{r}, & \gamma_{\theta z} &= \frac{1}{r}\frac{\partial u_z}{\partial \theta} + \frac{\partial u_\theta}{\partial z}, \\
\varepsilon_z &= \frac{\partial u_z}{\partial z}, & \gamma_{zr} &= \frac{\partial u_r}{\partial z} + \frac{\partial u_z}{\partial r}.
\end{aligned} \tag{1.30}$$

Similarly in the spherical coordinate system r, θ, φ, in which the displacement vector components are denoted by u_r, u_θ, u_φ we get the relationships

$$\varepsilon_r = \frac{\partial u_r}{\partial r},$$

$$\varepsilon_\theta = \frac{1}{r}\frac{\partial u_\theta}{\partial \theta} + \frac{u_r}{r},$$

$$\varepsilon_\varphi = \frac{1}{r\sin\theta}\frac{\partial u_\varphi}{\partial \theta} + \frac{u_\theta \operatorname{ctg}\theta}{r} + \frac{u_r}{r}, \qquad (1.31)$$

$$\gamma_{r\theta} = \frac{1}{r}\frac{\partial u_r}{\partial \theta} + \frac{\partial u_\theta}{\partial r} - \frac{u_\theta}{r},$$

$$\gamma_{\theta\varphi} = \frac{1}{r\sin\theta}\frac{\partial u_\theta}{\partial \varphi} + \frac{1}{r}\frac{\partial u_\varphi}{\partial \theta} - \frac{u_\varphi \operatorname{ctg}\varphi}{r},$$

$$\gamma_{\varphi r} = \frac{1}{r\sin\theta}\frac{\partial u_r}{\partial \varphi} + \frac{\partial u_\varphi}{\partial r} - \frac{u_\varphi}{r}.$$

By principal strains will be meant eigenvalues of the strain tensor ε_{ij}. These are calculated as the roots of the third order equation

$$\varepsilon^3 - J_{1\varepsilon}\varepsilon^2 + J_{2\varepsilon}\varepsilon - J_{3\varepsilon} = 0. \qquad (1.32)$$

$J_{1\varepsilon}, J_{2\varepsilon}, J_{3\varepsilon}$ are principal invariants of the strain tensor and can be presented as the homogeneous scalar functions of its components

$$\begin{aligned}
J_{1\varepsilon} &= \varepsilon_{ii} = \varepsilon_x + \varepsilon_y + \varepsilon_z, \\
J_{2\varepsilon} &= \tfrac{1}{2}(\varepsilon_{ii}\varepsilon_{jj} - \varepsilon_{ij}\varepsilon_{ji}) = \varepsilon_x\varepsilon_y + \varepsilon_y\varepsilon_z + \varepsilon_z\varepsilon_x - \tfrac{1}{4}(\gamma_{xy}^2 + \gamma_{yz}^2 + \gamma_{zx}^2), \\
J_{3\varepsilon} &= \tfrac{1}{6}e_{ijk}e_{pqr}\varepsilon_{ip}\varepsilon_{jq}\varepsilon_{kr} = \varepsilon_x\varepsilon_y\varepsilon_z + \tfrac{1}{4}\gamma_{xy}\gamma_{yz}\gamma_{zx} - \tfrac{1}{4}(\varepsilon_x\gamma_{yz}^2 + \varepsilon_y\gamma_{zx}^2 + \varepsilon_z\gamma_{xy}^2).
\end{aligned} \qquad (1.33)$$

The strain tensor can be written as the sum of the spherical $\varepsilon_m\delta_{ij}$ and deviatoric e_{ij} strains according to the equation

$$\varepsilon_{ij} = e_{ij} + \varepsilon_m\delta_{ij}, \qquad \text{where } \varepsilon_m = \tfrac{1}{3}\varepsilon_{kk}. \qquad (1.34)$$

Therefore the deviatoric strain has the form

$$e_{ij} = \begin{Bmatrix} \varepsilon_x - \varepsilon_m & \tfrac{1}{2}\gamma_{yx} & \tfrac{1}{2}\gamma_{zx} \\ \tfrac{1}{2}\gamma_{xy} & \varepsilon_y - \varepsilon_m & \tfrac{1}{2}\gamma_{zy} \\ \tfrac{1}{2}\gamma_{xz} & \tfrac{1}{2}\gamma_{yz} & \varepsilon_z - \varepsilon_m \end{Bmatrix}$$

$$= \begin{Bmatrix} \tfrac{2}{3}\varepsilon_x - \tfrac{1}{3}(\varepsilon_y + \varepsilon_z) & \tfrac{1}{2}\gamma_{yx} & \tfrac{1}{2}\gamma_{zx} \\ \tfrac{1}{2}\gamma_{xy} & \tfrac{2}{3}\varepsilon_y - \tfrac{1}{3}(\varepsilon_z + \varepsilon_x) & \tfrac{1}{2}\gamma_{zy} \\ \tfrac{1}{2}\gamma_{xz} & \tfrac{1}{2}\gamma_{yz} & \tfrac{2}{3}\varepsilon_z - \tfrac{1}{3}(\varepsilon_x + \varepsilon_y) \end{Bmatrix}. \qquad (1.35)$$

The principal invariants of the deviatoric strain J_{1e}, J_{2e}, J_{3e} are derived in a similar way to the invariants of deviatoric stress (1.18)

$$J_{1e} = e_{ii} = e_{xx} + e_{yy} + e_{zz} = 0,$$
$$J_{2e} = -\tfrac{1}{2}e_{ij}e_{ji} = -\tfrac{1}{2}[e_{xx}^2 + e_{yy}^2 + e_{zz}^2 + 2(e_{xy}^2 + e_{yz}^2 + e_{zx}^2)],$$
$$J_{3e} = \tfrac{1}{3}e_{ij}e_{jk}e_{ki} = \tfrac{1}{3}[e_{xx}^3 + e_{yy}^3 + e_{zz}^3 + 3(e_{xx}e_{xy}^2 + e_{xx}e_{xz}^2 \qquad (1.36)$$
$$+ e_{yy}e_{yz}^2 + e_{yy}e_{yx}^2 + e_{zz}e_{zx}^2 + e_{zz}e_{zy}^2) + 6e_{xy}e_{yz}e_{zx}],$$

and by using the strain tensor components ε_{ij} we get in engineering notation as an example

$$J_{2e} = -\tfrac{1}{6}[(\varepsilon_x - \varepsilon_y)^2 + (\varepsilon_y - \varepsilon_z)^2 + (\varepsilon_z - \varepsilon_x)^2 + \tfrac{3}{2}(\gamma_{xy}^2 + \gamma_{yz}^2 + \gamma_{zx}^2)],$$
$$J_{3e} = \tfrac{4}{9}\varepsilon_x\varepsilon_y\varepsilon_z - \tfrac{1}{9}(\varepsilon_x^2\varepsilon_y + \varepsilon_x^2\varepsilon_z + \varepsilon_y^2\varepsilon_x + \varepsilon_y^2\varepsilon_z + \varepsilon_z^2\varepsilon_x + \varepsilon_z^2\varepsilon_y) \qquad (1.37)$$
$$+ \tfrac{2}{27}(\varepsilon_x^3 + \varepsilon_y^3 + \varepsilon_z^3) - \tfrac{1}{6}(\varepsilon_x\gamma_{yz}^2 + \varepsilon_y\gamma_{zx}^2 + \varepsilon_z\gamma_{xy}^2) + \tfrac{1}{12}(\varepsilon_x\gamma_{xy}^2$$
$$+ \varepsilon_x\gamma_{xz}^2 + \varepsilon_y\gamma_{yx}^2 + \varepsilon_y\gamma_{yz}^2 + \varepsilon_z\gamma_{zx}^2 + \varepsilon_z\gamma_{zy}^2) + \tfrac{1}{4}\gamma_{xy}\gamma_{yz}\gamma_{zx}.$$

The invariants of the deviatoric strain J_{1e}, J_{2e}, J_{3e} can be expressed as functions of principal deviatoric strain components e_j, as well as by the invariants of the strain tensor $J_{1\varepsilon}, J_{2\varepsilon}, J_{3\varepsilon}$. These formulas have identical form to those given by equations (1.18) and (1.20) for the deviatoric stress invariants with formal replacement of the principal deviatoric stress components s_j by principal deviatoric strain components e_j. In addition to the strain tensor invariants, we define an effective strain $\bar{\varepsilon}$

$$\bar{\varepsilon} = \frac{\sqrt{2}}{2(1+\nu)}\sqrt{(\varepsilon_{xx} - \varepsilon_{yy})^2 + (\varepsilon_{yy} - \varepsilon_{zz})^2 + (\varepsilon_{zz} - \varepsilon_{xx})^2 + 6(\varepsilon_{xy}^2 + \varepsilon_{yz}^2 + \varepsilon_{zx}^2)},$$
$$\qquad (1.38)$$

where ν is a material constant called Poisson's ratio.

The effective strain so defined depends on the constant ν. But since this is inconvenient, we make the customary assumption that $\nu = 1/2$, leading to the following definition of effective strain

$$\bar{\varepsilon} = \sqrt{\tfrac{2}{3}e_{ij}e_{ij}} = \frac{2}{\sqrt{3}}\sqrt{-J_{2e}} \qquad (1.39)$$

$$= \frac{\sqrt{2}}{3}\sqrt{(\varepsilon_{xx} - \varepsilon_{yy})^2 + (\varepsilon_{yy} - \varepsilon_{zz})^2 + (\varepsilon_{zz} - \varepsilon_{xx})^2 + 6(\varepsilon_{xy}^2 + \varepsilon_{yz}^2 + \varepsilon_{zx}^2)}$$

$$= \tfrac{2}{3}\sqrt{\varepsilon_x^2 + \varepsilon_y^2 + \varepsilon_z^2 - (\varepsilon_x\varepsilon_y + \varepsilon_y\varepsilon_z + \varepsilon_z\varepsilon_x) + \tfrac{3}{4}(\gamma_{xy}^2 + \gamma_{yz}^2 + \gamma_{zx}^2)}.$$

The components of the infinitesimal strain tensor must fulfil the conditions of strain compatibility. By eliminating the displacement components u, v, w from the Cauchy equations (1.27) we get the set of six de Saint-Venant integrality equations

$$\frac{\partial^2 \varepsilon_{ij}}{\partial x_k \partial x_l} - \frac{\partial^2 \varepsilon_{ki}}{\partial x_j \partial x_l} - \frac{\partial^2 \varepsilon_{lj}}{\partial x_k \partial x_i} + \frac{\partial^2 \varepsilon_{kl}}{\partial x_j \partial x_i} = 0. \qquad (1.40)$$

In the Cartesian coordinate system this set of integrality equations takes the form

$$\frac{\partial^2 \varepsilon_{xx}}{\partial y^2} + \frac{\partial^2 \varepsilon_{yy}}{\partial x^2} - 2\frac{\partial^2 \varepsilon_{xy}}{\partial x \partial y} = 0,$$

$$\frac{\partial^2 \varepsilon_{yy}}{\partial z^2} + \frac{\partial^2 \varepsilon_{zz}}{\partial y^2} - 2\frac{\partial^2 \varepsilon_{yz}}{\partial y \partial z} = 0,$$

$$\frac{\partial^2 \varepsilon_{zz}}{\partial x^2} + \frac{\partial^2 \varepsilon_{xx}}{\partial z^2} - 2\frac{\partial^2 \varepsilon_{zx}}{\partial z \partial x} = 0, \qquad (1.41)$$

$$-\frac{\partial^2 \varepsilon_{xx}}{\partial y \partial z} + \frac{\partial}{\partial x}\left(-\frac{\partial \varepsilon_{yz}}{\partial x} + \frac{\partial \varepsilon_{zx}}{\partial y} + \frac{\partial \varepsilon_{xy}}{\partial z}\right) = 0,$$

$$-\frac{\partial^2 \varepsilon_{yy}}{\partial z \partial x} + \frac{\partial}{\partial y}\left(\frac{\partial \varepsilon_{yz}}{\partial x} - \frac{\partial \varepsilon_{zx}}{\partial y} + \frac{\partial \varepsilon_{xy}}{\partial z}\right) = 0,$$

$$-\frac{\partial^2 \varepsilon_{zz}}{\partial x \partial y} + \frac{\partial}{\partial z}\left(\frac{\partial \varepsilon_{yz}}{\partial x} + \frac{\partial \varepsilon_{zx}}{\partial y} - \frac{\partial \varepsilon_{xy}}{\partial z}\right) = 0.$$

In cylindrical coordinates r, θ, z a set of integrality equations has the form

$$\frac{2}{r}\frac{\partial^2 \varepsilon_{z\theta}}{\partial z \partial \theta} - \frac{1}{r^2}\frac{\partial^2 \varepsilon_{zz}}{\partial \theta^2} - \frac{\partial^2 \varepsilon_{\theta\theta}}{\partial z^2} + \frac{2}{r}\frac{\partial \varepsilon_{rz}}{\partial z} - \frac{1}{r}\frac{\partial \varepsilon_{zz}}{\partial r} = 0,$$

$$2\frac{\partial^2 \varepsilon_{rz}}{\partial r \partial z} - \frac{\partial^2 \varepsilon_{rr}}{\partial z^2} - \frac{\partial^2 \varepsilon_{zz}}{\partial r^2} = 0,$$

$$\frac{2\partial^2 \varepsilon_{\theta r}}{r \partial \theta \partial r} - \frac{\partial^2 \varepsilon_{\theta\theta}}{\partial r^2} - \frac{1}{r^2}\frac{\partial^2 \varepsilon_{rr}}{\partial \theta^2} + \frac{1}{r}\frac{\partial \varepsilon_{rr}}{\partial r} + \frac{2}{r^2}\frac{\partial \varepsilon_{r\theta}}{\partial \theta} - \frac{2}{r}\frac{\partial \varepsilon_{\theta\theta}}{\partial r} = 0,$$

$$\frac{1}{r}\frac{\partial^2 \varepsilon_{zz}}{\partial r \partial \theta} - \frac{\partial}{\partial z}\left(\frac{1}{r}\frac{\partial \varepsilon_{zr}}{\partial \theta} + \frac{\partial \varepsilon_{\theta z}}{\partial r} - \frac{\partial \varepsilon_{\theta r}}{\partial z}\right) + \frac{1}{r}\frac{\partial \varepsilon_{z\theta}}{\partial z} - \frac{1}{r^2}\frac{\partial \varepsilon_{zz}}{\partial \theta} = 0,$$

$$\frac{1}{r}\frac{\partial^2 \varepsilon_{rr}}{\partial \theta \partial z} - \frac{\partial}{\partial r}\left(\frac{1}{r}\frac{\partial \varepsilon_{zr}}{\partial \theta} - \frac{\partial \varepsilon_{\theta z}}{\partial r} + \frac{\partial \varepsilon_{\theta r}}{\partial z}\right) - \frac{2}{r}\frac{\partial \varepsilon_{r\theta}}{\partial z} + \frac{1}{r}\frac{\partial \varepsilon_{z\theta}}{\partial r} - \frac{1}{r^2}\frac{\partial \varepsilon_{z\theta}}{\partial z} = 0,$$

$$\frac{\partial^2 \varepsilon_{\theta\theta}}{\partial z \partial r} - \frac{1}{r}\frac{\partial}{\partial \theta}\left(-\frac{1}{r}\frac{\partial \varepsilon_{rz}}{\partial \theta} + \frac{\partial \varepsilon_{\theta z}}{\partial r} + \frac{\partial \varepsilon_{\theta r}}{\partial z}\right) + \frac{2}{r}\frac{\partial \varepsilon_{zr}}{\partial r}$$

$$-\frac{1}{r^2}\frac{\partial \varepsilon_{\theta z}}{\partial \theta} - \frac{1}{r}\frac{\partial}{\partial z}(\varepsilon_{rr} - \varepsilon_{zz}) = 0.$$

2 Finite Strains

The majority of problems considered in this book are based on the assumption of infinitesimally small strains. However in some problems it is necessary to deviate from this assumption. In this chapter we introduce elementary information and relationships in the realm of finite strains.

2.1 Finite Strain Tensor in Material and Spatial Descriptions

Let B_{t_0} denote a domain occupied by the body in physical space at initial time t_0, and B_t at current time t. Let

$$x = x(X) \tag{2.1}$$

denote deformation from the reference configuration B of the Euclidean point space E to the current one. Introduce bases G_I and g_i in the reference and current configurations respectively, so that in component form (rectangular Cartesian) Eq. (2.1) may be written as

$$x_i = x_i(X_I) \qquad i, I = 1, 2, 3. \tag{2.2}$$

One and the same elementary particle P occupies in initial state the point with Cartesian coordinates $X_I(X, Y, Z)$, and in the current time $x_i(x, y, z)$. Coordinates X, Y, Z treated as independent variables are called material coordinates, and coordinates x, y, z assumed as independent variables are called spatial coordinates. The notion of material coordinates (also called the Lagrange coordinates) was introduced by Euler, and the notion of spatial coordinates (sometimes called Euler coordinates) was introduced by d'Alembert (1752). Using material and spatial descriptions the motion of the body can be represented by the relations

$$x_i = x_i(X_k, t) \quad \text{or} \quad X_I = X_I(x_k, t) \tag{2.3}$$

respectively.

The tensor F defined by

$$F = \text{grad } x(X) = \frac{\partial x(X)}{\partial X} = \frac{\partial x}{\partial X} \tag{2.4}$$

is termed the deformation gradient and is a two point tensor (Ericksen 1960). The Jacobian of the mapping (2.1) is given by

$$J = \det F. \tag{2.5}$$

According to the polar decomposition theorem the tensor F can be represented uniquely by

$$F = RU = VR \tag{2.6}$$

where U, V are positive–definite, symmetric tensors and R is an orthogonal tensor. Tensors U and V are termed the right and left stretch tensor respectively. They have common eigenvalues λ_α ($\alpha = 1, 2, 3$) called principal stretches and corresponding eigenvectors N^α and n^α known as the Lagrangian and Eulerian triads respectively. Note that

$$n^\alpha = RN^\alpha, \tag{2.7}$$

i.e. the deformation rotates eigenvectors of U into those of V. In a majority of analyses one assumes the Euclidean metric tensors which become identical and equal to the Kronecker delta δ_{ij} in the Cartesian coordinate system. For simplicity we consider henceforward finite deformations in space with Euclidean metric tensors. At the end of the chapter we present the basic principles of finite strains with the introduction of arbitrary base vectors.

Explanation of notions of finite deformation should be carried out in some detail. Let us start from the tensoral strain measures in space with Euclidean metric tensors. In the Cartesian coordinate system the displacement vector $P_0 P_1$ (Fig. 2.1) has material components $U_I(U, V, W)$ or spatial components $u_i(u, v, w)$.

$$U_I = x_i(X_k, t) - X_I \quad \text{or} \quad u_i = x_i - X_I(x_k, t). \tag{2.8}$$

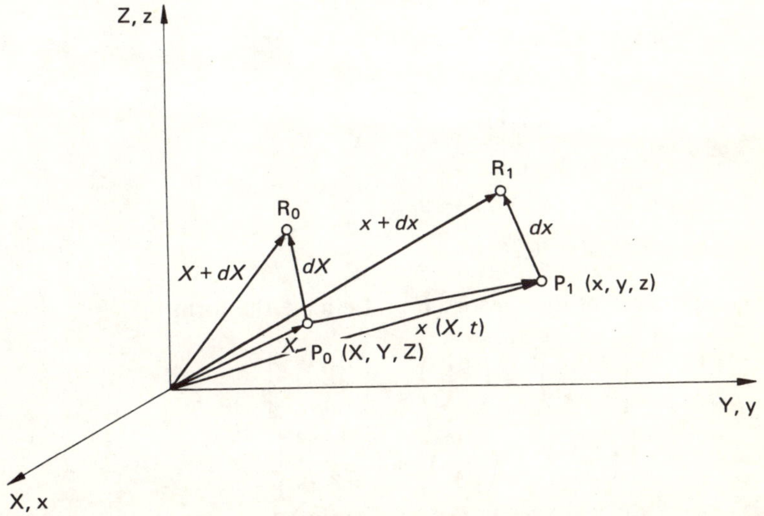

Figure 2.1. Material coordinates (Lagrangian) and spatial coordinates (Eulerian); displacement vector.

In order to introduce strain measures in both of the descriptions discussed we consider besides the particle P a second material particle R placed 'near' to it. The square of the magnitude of vector $PR = dX$ in the initial configuration is $dS^2 = dX_I dX_I$, and in the current configuration $ds^2 = dx_i dx_i$. Making use of the relationships between differentials dx_i, dX_I which follow from the motion law (2.8),

$$dx_i = \frac{\partial x_i}{\partial X_k} dX_k \quad \text{or} \quad dX_I = \frac{\partial X_I}{\partial x_k} dx_k \tag{2.9}$$

we get

$$ds^2 - dS^2 = dx_i dx_i - dX_I dX_I = dx_i dx_i - dX_K dX_M \delta_{KM}$$
$$= (x_{i,K} x_{i,M} - \delta_{KM}) dX_K dX_M = 2 E_{KM} dX_K dX_M \tag{2.10}$$

or

$$ds^2 - dS^2 = dx_k dx_m \delta_{km} - dX_I dX_I = (\delta_{km} - X_{I,k} X_{I,m}) dx_k dx_m$$
$$= 2 e_{km} dx_k dx_m, \tag{2.11}$$

where E_{KM} is the Green strain tensor, also called the Green–Lagrange strain tensor, and e_{km} is the Almansi strain tensor. Expressing next the material deformation gradients in (2.10) $X_{i,K}$ and $x_{i,M}$ as well as the spatial deformation gradients in (2.11) $X_{I,k}$ and $X_{I,m}$ in terms of displacements given in (2.8) we get expressions for the two tensors

$$E_{KM} = \frac{1}{2} \left[\frac{\partial x_i}{\partial X_K} \frac{\partial x_i}{\partial X_M} - \delta_{KM} \right] = \tfrac{1}{2}(x_{i,K} x_{i,M} - \delta_{KM})$$
$$= \tfrac{1}{2}[(U_{I,K} + \delta_{IK})(U_{I,M} + \delta_{IM}) - \delta_{KM}] \tag{2.12}$$
$$= \tfrac{1}{2}(U_{M,K} + U_{K,M} + U_{I,K} U_{I,M}),$$

and

$$e_{km} = \frac{1}{2} \left[\delta_{km} - \frac{\partial X_I}{\partial x_k} \frac{\partial X_I}{\partial x_m} \right] = \tfrac{1}{2}(\delta_{km} - X_{I,k} X_{I,m})$$
$$= \tfrac{1}{2}[\delta_{km} - (\delta_{ik} - u_{i,k})(\delta_{im} - u_{i,m})] \tag{2.13}$$
$$= \tfrac{1}{2}(u_{m,k} + u_{k,m} - u_{i,k} u_{i,m}).$$

The components of the tensors E and e have the form

$$E_{XX} = \frac{\partial U}{\partial X} + \frac{1}{2}\left[\left(\frac{\partial U}{\partial X}\right)^2 + \left(\frac{\partial V}{\partial X}\right)^2 + \left(\frac{\partial W}{\partial X}\right)^2\right],$$

$$E_{YY} = \frac{\partial V}{\partial Y} + \frac{1}{2}\left[\left(\frac{\partial U}{\partial Y}\right)^2 + \left(\frac{\partial V}{\partial Y}\right)^2 + \left(\frac{\partial W}{\partial Y}\right)^2\right],$$

$$E_{ZZ} = \frac{\partial W}{\partial Z} + \frac{1}{2}\left[\left(\frac{\partial U}{\partial Z}\right)^2 + \left(\frac{\partial V}{\partial Z}\right)^2 + \left(\frac{\partial W}{\partial Z}\right)^2\right],$$

$$E_{XY} = \frac{1}{2}\left[\frac{\partial U}{\partial Y} + \frac{\partial V}{\partial X} + \frac{\partial U}{\partial X}\frac{\partial U}{\partial Y} + \frac{\partial V}{\partial X}\frac{\partial V}{\partial Y} + \frac{\partial W}{\partial X}\frac{\partial W}{\partial Y}\right],$$

$$E_{YZ} = \frac{1}{2}\left[\frac{\partial V}{\partial Z} + \frac{\partial W}{\partial Y} + \frac{\partial U}{\partial Y}\frac{\partial U}{\partial Z} + \frac{\partial V}{\partial Y}\frac{\partial V}{\partial Z} + \frac{\partial W}{\partial Y}\frac{\partial W}{\partial Z}\right],$$

$$E_{ZX} = \frac{1}{2}\left[\frac{\partial W}{\partial X} + \frac{\partial U}{\partial Z} + \frac{\partial U}{\partial Z}\frac{\partial U}{\partial X} + \frac{\partial V}{\partial Z}\frac{\partial V}{\partial X} + \frac{\partial W}{\partial Z}\frac{\partial W}{\partial X}\right],$$

(2.14)

$$e_{xx} = \frac{\partial u}{\partial x} - \frac{1}{2}\left[\left(\frac{\partial u}{\partial x}\right)^2 + \left(\frac{\partial v}{\partial x}\right)^2 + \left(\frac{\partial w}{\partial x}\right)^2\right],$$

$$e_{yy} = \frac{\partial v}{\partial y} - \frac{1}{2}\left[\left(\frac{\partial u}{\partial y}\right)^2 + \left(\frac{\partial v}{\partial y}\right)^2 + \left(\frac{\partial w}{\partial y}\right)^2\right],$$

$$e_{zz} = \frac{\partial w}{\partial z} - \frac{1}{2}\left[\left(\frac{\partial u}{\partial z}\right)^2 + \left(\frac{\partial v}{\partial z}\right)^2 + \left(\frac{\partial w}{\partial z}\right)^2\right],$$

$$e_{xy} = \frac{1}{2}\left[\frac{\partial u}{\partial y} + \frac{\partial v}{\partial x} - \frac{\partial u}{\partial x}\frac{\partial u}{\partial y} - \frac{\partial v}{\partial x}\frac{\partial v}{\partial y} - \frac{\partial w}{\partial x}\frac{\partial w}{\partial y}\right],$$

$$e_{yz} = \frac{1}{2}\left[\frac{\partial v}{\partial z} + \frac{\partial w}{\partial y} - \frac{\partial u}{\partial y}\frac{\partial u}{\partial z} - \frac{\partial v}{\partial y}\frac{\partial v}{\partial z} - \frac{\partial w}{\partial y}\frac{\partial w}{\partial z}\right],$$

$$e_{zx} = \frac{1}{2}\left[\frac{\partial w}{\partial x} + \frac{\partial u}{\partial z} - \frac{\partial u}{\partial z}\frac{\partial u}{\partial x} - \frac{\partial v}{\partial z}\frac{\partial v}{\partial x} - \frac{\partial w}{\partial z}\frac{\partial w}{\partial x}\right].$$

(2.15)

In the case of small displacements one can omit the products in (2.15); then, neglecting differentiation between material and spatial coordinates, we get both from (2.12) as well as from (2.13) identical expressions for the infinitesimal strain tensor

$$e_{km} = \tfrac{1}{2}(u_{k,m} + u_{m,k}). \tag{2.16}$$

In an arbitrary curvilinear coordinate system Eqs. (2.9), (2.12) and (2.13) have to be changed. The square of the magnitude of vectors $d\mathbf{X}$ and $d\mathbf{x}$ are given as

$$dS^2 = G_{KM}dX^K dX^M, \qquad ds^2 = g_{km}dx^k dx^m, \tag{2.17}$$

and

$$dS^2 = c_{km}dx^k dx^m, \qquad ds^2 = C_{KM}dX^K dX^M, \tag{2.18}$$

where G_{KM} and g_{km} are metric tensors for the coordinate system X_I or x_i, and C_{KM} and c_{km} and denote Cauchy–Green strain tensors. In place of the definitions (2.12) and (2.13) we then get

$$E_{KM} = \tfrac{1}{2}(C_{KM} - G_{KM}) = \tfrac{1}{2}(g_{km}x^k_{,K}x^m_{,M} - G_{KM}),$$
$$e_{km} = \tfrac{1}{2}(g_{km} - c_{km}) = \tfrac{1}{2}(g_{km} - G_{KM}X^K_{,k}X^M_{,m}).$$

(2.19)

In the case of an orthogonal curvilinear coordinate system the components of the metric tensors G_{KM} and g_{km} are expressed as

$$G_{KM} = \begin{cases} H_K^2 & (K = M), \\ 0 & (K \neq M), \end{cases} \qquad g_{km} = \begin{cases} h_k^2 & (k = m), \\ 0 & (k \neq m) \end{cases} \tag{2.20}$$

where H_K and h_k are material parameters. In cylindrical coordinates r, θ, z by using the respective transformation rules to the Cartesian coordinates x, y, z

$$x = r\cos\theta, \qquad y = r\sin\theta, \qquad z = Z, \tag{2.21}$$

we get for the square of the element length ds

$$ds^2 = dx^2 + dy^2 + dz^2 = 1\,dr^2 + r^2\,d\theta + 1\,dz^2. \tag{2.22}$$

In the initial configuration

$$dS^2 = dX^2 + dY^2 + dZ^2 = 1\,dR^2 + R^2\,d\theta + 1\,dZ^2. \tag{2.23}$$

The metric tensors G_{KM} and g_{km} then have the form

$$G_{KM} = \begin{cases} 1 & 0 & 0 \\ 0 & R^2 & 0 \\ 0 & 0 & 1 \end{cases}, \qquad g_{km} = \begin{cases} 1 & 0 & 0 \\ 0 & r^2 & 0 \\ 0 & 0 & 1 \end{cases}. \tag{2.24}$$

The components of the strain tensor E in the cylindrical coordinate system $r = r(R, \Theta, Z)$, $\theta = \theta(R, \Theta, Z)$, $z = z(R, \Theta, Z)$ are of the form

$$\begin{aligned}
E_{RR} &= \frac{1}{2}\left[\left(\frac{\partial r}{\partial R}\right)^2 + r^2\left(\frac{\partial \theta}{\partial R}\right)^2 + \left(\frac{\partial z}{\partial R}\right)^2 - 1\right], \\
E_{\Theta\Theta} &= \frac{1}{2}\left[\left(\frac{\partial r}{\partial \Theta}\right)^2 + r^2\left(\frac{\partial \theta}{\partial \Theta}\right)^2 + \left(\frac{\partial z}{\partial \Theta}\right)^2 - R^2\right], \\
E_{ZZ} &= \frac{1}{2}\left[\left(\frac{\partial r}{\partial Z}\right)^2 + r^2\left(\frac{\partial \theta}{\partial Z}\right)^2 + \left(\frac{\partial z}{\partial Z}\right)^2 - 1\right], \\
E_{R\Theta} &= E_{\Theta R} = \frac{1}{2}\left[\frac{\partial r}{\partial R}\frac{\partial r}{\partial \Theta} + r^2\frac{\partial \theta}{\partial R}\frac{\partial \theta}{\partial \Theta} + \frac{\partial z}{\partial R}\frac{\partial z}{\partial \Theta}\right], \\
E_{\Theta Z} &= E_{Z\Theta} = \frac{1}{2}\left[\frac{\partial r}{\partial \Theta}\frac{\partial r}{\partial Z} + r^2\frac{\partial \theta}{\partial \Theta}\frac{\partial \theta}{\partial Z} + \frac{\partial z}{\partial \Theta}\frac{\partial z}{\partial Z}\right], \\
E_{ZR} &= E_{RZ} = \frac{1}{2}\left[\frac{\partial r}{\partial Z}\frac{\partial r}{\partial R} + r^2\frac{\partial \theta}{\partial Z}\frac{\partial \theta}{\partial R} + \frac{\partial z}{\partial Z}\frac{\partial z}{\partial R}\right].
\end{aligned} \tag{2.25}$$

In the particular case of plane deformation where

$$r = R + U(R, \Theta), \qquad \theta = \Theta + \beta(R, \Theta), \qquad z = Z, \tag{2.26}$$

non-zero components of the tensor E are

$$E_{RR} = \frac{\partial U}{\partial R} + \frac{1}{2}\left(\frac{\partial U}{\partial R}\right)^2 + \frac{1}{2}(R+U)^2\left(\frac{\partial \beta}{\partial R}\right)^2,$$

$$E_{\Theta\Theta} = RU + R^2\frac{\partial \beta}{\partial \Theta} + \frac{1}{2}U^2 + 2RU\frac{\partial \beta}{\partial \Theta} + U^2\frac{\partial \beta}{\partial \Theta} + \frac{1}{2}\left(\frac{\partial U}{\partial \Theta}\right)^2$$

$$+ \frac{1}{2}(R+U)^2\left(\frac{\partial \beta}{\partial \Theta}\right)^2, \tag{2.27}$$

$$E_{R\Theta} = E_{\Theta R} = \frac{1}{2}\left[\frac{\partial U}{\partial \Theta} + R^2\frac{\partial \beta}{\partial R} + \frac{\partial U}{\partial R}\frac{\partial U}{\partial \Theta} + 2RU\frac{\partial \beta}{\partial R}\right.$$

$$\left. + U^2\frac{\partial \beta}{\partial R} + (R+U)^2\frac{\partial \beta}{\partial R}\frac{\partial \beta}{\partial \Theta}\right].$$

In another important case of axi-symmetric deformation with respect to the Z axis we get for the deformation describing the combined torsion and extension of a cylindrical tube the following relations

$$r = R + U(R), \qquad \theta = \Theta + \lambda\psi Z, \qquad z = \lambda Z \tag{2.28}$$

and

$$E_{KM} = \begin{Bmatrix} \frac{\partial U}{\partial R} + \frac{1}{2}\left(\frac{\partial U}{\partial R}\right)^2 & 0 & 0 \\ 0 & RU + \frac{1}{2}U^2 & \frac{1}{2}(R+U)^2\lambda\psi \\ 0 & \frac{1}{2}(R+U)^2\lambda\psi & \frac{1}{2}[\lambda^2 - 1 + (R+U)^2\lambda^2\psi^2] \end{Bmatrix}. \tag{2.29}$$

In the above relations ψ is the angle of torsion on unit length in the deformed state and λ is a nondimensional extension coefficient. Appearing in Eqs (2.25)–(2.29) tensor components E_{KM} in curvilinear coordinate systems have various dimensions. Practically, application to physical equations requires definition of the so-called physical components $E_{(KM)}$. They possess the same dimensions, but do not transform in accordance with the tensor transformation rule so are not components of the tensor. Determination of physical components is simple only in the case of orthogonal curvilinear coordinates, when

$$E_{(KM)} = \frac{E_{KM}}{H_K H_M}, \tag{2.30}$$

where H_K and H_M are material constants. Then in the problem considered of extension with torsion, physical components of the $E_{(KM)}$ tensor are expressed as

$$E_{(KM)} = \begin{Bmatrix} \dfrac{\partial U}{\partial R} + \dfrac{1}{2}\left(\dfrac{\partial U}{\partial R}\right)^2 & 0 & 0 \\ 0 & \dfrac{U}{R} + \dfrac{1}{2}\left(\dfrac{U}{R}\right)^2 & \dfrac{1}{2}\left(1 + \dfrac{U}{R}\right)^2 R\lambda\psi \\ 0 & \dfrac{1}{2}\left(1 + \dfrac{U}{R}\right)^2 R\lambda\psi & \dfrac{1}{2}\left[\lambda^2 - 1 + \left(1 + \dfrac{U}{R}\right)^2 R^2\lambda^2\psi^2\right] \end{Bmatrix}. \tag{2.31}$$

On the basis of finite strain tensors E and e Truesdell and Toupin (1960) define logarithmic strain tensors in material and spatial descriptions

$$H = \tfrac{1}{2}\ln(1 + 2E) \quad \text{or} \quad h = -\tfrac{1}{2}\ln(1 - 2e). \tag{2.32}$$

Definitions (2.32) should be understood as expansions in infinite series of logarithms or their analytical extensions

$$\begin{aligned} H &= \frac{1}{2}\left[2E - \frac{(2E)^2}{2} + \frac{(2E)^3}{3} - \cdots\right], \\ h &= -\frac{1}{2}\left[-2e - \frac{(2e)^2}{2} - \frac{(2e)^2}{3} - \cdots\right]. \end{aligned} \tag{2.33}$$

Using orthogonal Cartesian coordinates we get respectively

$$\begin{aligned} H_{KM} &= \frac{1}{2}\left[2E_{KM} - \frac{2^2}{2} E_{KI} E_{IM} + \frac{2^3}{3} E_{KI} E_{IJ} E_{JM} - \cdots\right], \\ h_{km} &= \frac{1}{2}\left[2e_{km} + \frac{2^2}{2} e_{ki} e_{im} + \frac{2^3}{3} e_{ki} e_{ij} e_{jm} + \cdots\right]. \end{aligned} \tag{2.34}$$

In the general case of deformation, the use of the definitions (2.32)–(2.34) is complicated because of the necessity for repeated multiplication of the tensors E^α and e^α by each other (with simultaneous contraction) and the use of infinite series. Only in the case of nonrotational deformation, when the principal directions are materially constant, can we easily ascertain that the definitions of both tensors H and h have identical form, and then the series can be written in finite form. In this case the principal components correspond to the logarithmic Hencky strains ε_α^L

$$\varepsilon_\alpha^L = \varepsilon_\alpha^H = \varepsilon_\alpha^h = \ln \lambda_\alpha \qquad (\alpha = 1, 2, 3) \tag{2.35}$$

where $\lambda_\alpha = (dx/dX)_\alpha$ are principal stretches ($0 < \lambda_\alpha < \alpha$). In the case of nonrotational deformation all the above strain measures can be obtained on the basis of the following definition (Seth, 1962)

$$\varepsilon_\alpha = \frac{1}{n}[1 - (\lambda_\alpha)^{-n}]. \tag{2.36}$$

For various values of the index n we get

$$\begin{aligned}
n &= -1, & \varepsilon_\alpha &= \varepsilon_\alpha^B = \lambda_\alpha - 1 & (-1 &< \varepsilon_\alpha^B < \infty), \\
n &= -2, & \varepsilon_\alpha &= \varepsilon_\alpha^E = \tfrac{1}{2}(\lambda_\alpha^2 - 1) & (-\tfrac{1}{2} &< \varepsilon_\alpha^E < \infty), \\
n &= 2, & \varepsilon_\alpha &= \varepsilon_\alpha^e = \tfrac{1}{2}(1 - \lambda_\alpha^{-2}) & (-\infty &< \varepsilon_\alpha^e < \tfrac{1}{2}), \\
n &= 0, & \varepsilon_\alpha &= \varepsilon_\alpha^L = \ln \lambda_\alpha & (-\infty &< \varepsilon_\alpha^L < \infty)
\end{aligned} \qquad (2.37)$$

where $\varepsilon_\alpha^B, \varepsilon_\alpha^E, \varepsilon_\alpha^e$ denote principal components of the Biot, Green–Lagrange and Almansi strain tensors respectively. The above comparison shows that of the strain measures discussed only the logarithmic measure has properties of additive symmetry in tension–compression

$$f(1/\lambda) = -f(\lambda). \qquad (2.38)$$

The elongation λ indicates symmetry of multiplicative type. The other strain measures are not symmetric ones at all. In the case of nonrotational axisymmetric deformation, putting $\psi = 0$ in Eq. (2.31) and making use of definition (2.32), we get

$$H_{KM} = \begin{Bmatrix} \ln\left(1 + \dfrac{dU}{dR}\right) & 0 & 0 \\ 0 & \ln\left(1 + \dfrac{U}{R}\right) & 0 \\ 0 & 0 & \ln \lambda \end{Bmatrix}. \qquad (2.39)$$

In the case $\psi \neq 0$ the tensor components H_{KM} are expressed by infinite series (2.33), (2.34).

2.2 Deformation Rate Tensor

The velocity v_i of the particle X_K is defined as changes of particle position with time. In the material description we have

$$v_i \stackrel{\text{def}}{=} \left[\frac{\partial x_i(X_K, t)}{\partial t}\right]_{X = \text{const}} = v_i(X_K, t) \qquad (2.40)$$

and substituting in place of the motion $X_K = X_K(x_m, t)$ we get the velocity field in the spatial description

$$v_i = v_i[X_K(x_m, t), t] = v_i(x_m, t). \qquad (2.41)$$

The acceleration a_i of the particle X_K is defined as changes of the velocity of the particle with time

$$a_i \stackrel{\text{def}}{=} \left[\frac{\partial v_i(X_K, t)}{\partial t}\right]_{X = \text{const}}. \qquad (2.42)$$

This magnitude should not be confused with the derivative $[\partial v_i/\partial t]_{x = \text{const}}$. The partial derivative of an arbitrary function f with respect to time t depends on the description assumed: material $f(X, t)$ or spatial $f(x, t)$ denote two different magnitudes. In the first case $[\partial f(X, t)/\partial t]$ describes the changes of function f for the fixed particle X during the time of its motion. In the second case

$[\partial f(x, t)/\partial t]_{x = \text{const}}$ denotes changes of the function f at the fixed point x of the space and refers to two different particles, which in two different moments t and $t + dt$ (where dt is small) occupy the same position x. For an arbitrary function $F(X_K, t)$ given in Lagrangian Cartesian coordinates we get

$$\dot{F}(X_K, t) = \frac{DF(X_K, t)}{Dt} = \left[\frac{\partial F(X_K, t)}{\partial t}\right]_X. \tag{2.43}$$

For a function $f(x_k, t)$ given in Eulerian Cartesian coordinates the material derivative is calculated by the formula

$$\dot{f}(x_k, t) = \frac{Df(x_k, t)}{Dt} = \left[\frac{\partial f(x_k, t)}{\partial t}\right]_x + \frac{\partial f(x_k, t)}{\partial x_m}\left[\frac{\partial x_m(x_k, t)}{\partial t}\right]_X$$

$$= \left(\frac{\partial f}{\partial t}\right)_x + \frac{\partial f}{\partial x_m} v_m. \tag{2.44}$$

In order to define the deformation rate tensor one calculates the material derivative $\frac{D}{Dt}$ of each side of the expression on square of the material fibre length, represented in time t by the vector dx, $(ds)^2 = dx_i dx_i$. Considering the relations

$$\frac{D}{Dt}(dx_i) = \frac{D}{Dt}\left(\frac{\partial x_i}{\partial X_K} dX_K\right) = \frac{\partial v_i(X_K, t)}{\partial X_K} dX_K = \frac{\partial v_i(x_k, t)}{\partial x_k} dx_k \tag{2.45}$$

we get

$$\frac{D}{Dt}(ds)^2 = 2v_{i,j}(x, t) dx_i dx_j.$$

Writing

$$v_{i,j} = d_{ij} + \omega_{ij},$$

where

$$d_{ij} = \tfrac{1}{2}[v_{i,j}(x, t) + v_{j,i}(x, t)] \quad (d_{ij} = d_{ji}), \tag{2.46}$$

and

$$\omega_{ij} = \tfrac{1}{2}[v_{i,j}(x, t) - v_{j,i}(x, t)] \quad (\omega_{ij} = -\omega_{ji}),$$

we get

$$\frac{D}{Dt}(ds)^2 = 2(d_{ij} + \omega_{ij}) dx_i dx_j = 2d_{ij} dx_i dx_j, \tag{2.47}$$

because the product of the skew-symmetric tensor ω_{ij} and symmetric tensor d_{ij} is equal to zero.

The tensor d_{ij} (symmetric part of the spatial velocity gradient) is called the deformation rate tensor. This tensor resembles in its construction the infinitesimal strain rate tensor ε_{IJ}

$$\frac{d\varepsilon_{IJ}}{dt} = \frac{1}{2}\left(\frac{\partial v_I}{\partial x_J} + \frac{\partial v_J}{\partial x_I}\right). \tag{2.48}$$

However in the tensor d_{ij} the derivatives relative to coordinates x_i appear, whereas in the tensor $\dfrac{d\varepsilon_{IJ}}{dt}$ derivatives are relative to the material coordinates X_I.

Tensor d_{ij} has clear physical interpretation. The deformation rate components d_{ij} are measures of instantaneous velocity changes in the length and angle of the material element (in time t). In particular, material derivatives of principal logarithmic strains are the same as the normal components of the deformation rate tensor in its principal directions

$$(\ln \lambda_\alpha)^{\cdot} = \frac{\dot{\lambda}_\alpha}{\lambda_\alpha} = d_\alpha \qquad (\alpha = 1, 2, 3). \tag{2.49}$$

Note particularly the difference between the two notations: effective strain rate $\dot{\bar{\varepsilon}}$, and the derivative of $\bar{\varepsilon}$ with respect to time, $(\bar{\varepsilon})^{\cdot}$. The effective strain rate is an invariant of the strain rate tensor $\dot{\varepsilon}_{IJ}$ and is defined as

$$\dot{\bar{\varepsilon}} \stackrel{\text{def}}{=} \frac{2}{\sqrt{3}} \sqrt{-J_{2\dot{\varepsilon}}} = \sqrt{\tfrac{2}{3} \dot{e}_{IJ} \dot{e}_{IJ}}, \tag{2.50}$$

and

$$(\bar{\varepsilon})^{\cdot} \stackrel{\text{def}}{=} \left[\frac{2}{\sqrt{3}} \sqrt{-J_{2e}}\right]^{\cdot} = \left[\sqrt{\tfrac{2}{3} e_{IJ} e_{IJ}}\right]^{\cdot}. \tag{2.51}$$

Only in some special cases of deformation, e.g. uniaxial extension with plane strain and volume-preserving deformation, are the two definitions identical.

In cylindrical coordinates r, θ, z, physical components of the gradient of the velocity vector and the deformation rate tensor have the following representation

$$v_{i,j} = \left\{ \begin{array}{ccc} \dfrac{\partial v_r}{\partial r} & \dfrac{1}{r}\dfrac{\partial v_r}{\partial \theta} - \dfrac{v_\theta}{r} & \dfrac{\partial v_r}{\partial z} \\ \dfrac{\partial v_\theta}{\partial r} & \dfrac{1}{r}\dfrac{\partial v_\theta}{\partial \theta} + \dfrac{v_r}{r} & \dfrac{\partial v_\theta}{\partial z} \\ \dfrac{\partial v_z}{\partial r} & \dfrac{1}{r}\dfrac{\partial v_z}{\partial \theta} & \dfrac{\partial v_z}{\partial z} \end{array} \right\} \tag{2.52}$$

and

$$d_{rr} = \frac{\partial v_r}{\partial r}, \qquad d_{\theta\theta} = \frac{1}{r}\frac{\partial v_\theta}{\partial \theta} + \frac{v_r}{r}, \qquad d_{zz} = \frac{\partial v_z}{\partial z},$$

$$d_{r\theta} = d_{\theta r} = \frac{1}{2}\left(\frac{1}{r}\frac{\partial v_r}{\partial \theta} + \frac{\partial v_\theta}{\partial r} - \frac{v_\theta}{r}\right),$$

$$d_{\theta z} = d_{z\theta} = \frac{1}{2}\left(\frac{1}{r}\frac{\partial v_z}{\partial \theta} + \frac{\partial v_\theta}{\partial z}\right), \tag{2.53}$$

$$d_{rz} = d_{zr} = \frac{1}{2}\left(\frac{\partial v_r}{\partial z} + \frac{\partial v_z}{\partial r}\right).$$

In the spherical coordinate system r, θ, φ, physical components of the velocity gradient and strain rate tensor may be written

$$v_{i,j} = \begin{Bmatrix} \dfrac{\partial v_r}{\partial r} & \dfrac{1}{r}\dfrac{\partial v_r}{\partial \theta} - \dfrac{v_\theta}{r} & \dfrac{1}{r\sin\theta}\dfrac{\partial v_r}{\partial \varphi} - \dfrac{v_\varphi}{r} \\ \dfrac{\partial v_\theta}{\partial r} & \dfrac{1}{r}\dfrac{\partial v_\theta}{\partial \theta} + \dfrac{v_r}{r} & \dfrac{1}{r\sin\theta}\dfrac{\partial v_\theta}{\partial \varphi} - \dfrac{v_\varphi}{r}\operatorname{ctg}\varphi \\ \dfrac{\partial v_\varphi}{\partial r} & \dfrac{1}{r}\dfrac{\partial v_\varphi}{\partial \theta} & \dfrac{1}{r\sin\theta}\dfrac{\partial v_\varphi}{\partial \varphi} + \dfrac{v_\theta}{r}\operatorname{ctg}\theta + \dfrac{v_r}{r} \end{Bmatrix} \quad (2.54)$$

and

$$d_{rr} = \frac{\partial v_r}{\partial r}, \qquad d_{\theta\theta} = \frac{1}{r}\frac{\partial v_\theta}{\partial \theta} + \frac{v_r}{r},$$

$$d_{\varphi\varphi} = \frac{1}{r\sin\theta}\frac{\partial v_\varphi}{\partial \varphi} + \frac{v_\theta}{r}\operatorname{ctg}\theta + \frac{v_r}{r},$$

$$d_{r\theta} = d_{\theta r} = \frac{1}{2}\left(\frac{1}{r}\frac{\partial v_r}{\partial \theta} + \frac{\partial v_\theta}{\partial r} - \frac{v_\theta}{r}\right), \quad (2.55)$$

$$d_{\theta\varphi} = d_{\varphi\theta} = \frac{1}{2}\left(\frac{1}{r\sin\theta}\frac{\partial v_\theta}{\partial \varphi} + \frac{1}{r}\frac{\partial v_\varphi}{\partial \theta} - \frac{v_\varphi}{r}\operatorname{ctg}\varphi\right),$$

$$d_{\varphi r} = d_{r\varphi} = \frac{1}{2}\left(\frac{1}{r\sin\theta}\frac{\partial v_r}{\partial \varphi} + \frac{\partial v_\varphi}{\partial r} - \frac{v_\varphi}{r}\right).$$

The formulas (2.53) and (2.55) have the same construction as the formulas (1.30) and (1.31) but with the displacement components replaced by velocity components. However Eqs. (1.30) and (1.31) are formulated in the material coordinate system R, Θ, Z or R, Θ, Φ and Eqs. (2.53), (2.55) in spatial coordinates r, θ, z or r, θ, ϕ.

Besides the deformation rate tensor $d(x, t)$ we introduce also the Green–Lagrange strain rate tensor $\dot{E}(X, t)$, calculated as the material derivative of the Green–Lagrange finite strain. Calculating the material derivative from the expression of the square of the material fibre length represented in the undeformed state by the vector dX, we get

$$\frac{D}{Dt}(ds)^2 = 2\frac{D}{Dt}[E_{KM}(X,t)]dX_K dX_M. \quad (2.56)$$

The tensors $d(x, t)$ and $\dot{E}(X, t)$ are different measures of strain rate in Euler and Lagrange formulation, respectively. They are connected by the relationship

$$\dot{E}_{KM}(X,t) = \frac{DE_{KM}}{Dt} = \left[\frac{\partial x_i}{\partial X_K}\frac{\partial x_j}{\partial X_M}d_{ij}(x,t)\right]_{x=x(X,t)}. \quad (2.57)$$

All of the above discussed strain tensors can be expressed in the curvilinear coordinate system by the general Lagrangian strain tensor

$$\tilde{E} = \begin{cases} \frac{1}{m}(U^m - 1), & m \neq 0 \\ \ln U, & m = 0 \end{cases} \tag{2.58}$$

and general Eulerian strain tensor

$$\tilde{e} = \begin{cases} \frac{1}{m}(V^m - 1), & m \neq 0 \\ \ln V, & m = 0 \end{cases} \tag{2.59}$$

with eigenvalues

$$f(\lambda_\alpha) = \begin{cases} \frac{1}{m}(\lambda_\alpha^m - 1), & m \neq 0 \\ \ln \lambda_\alpha, & m = 0 \end{cases} \tag{2.60}$$

($\alpha = 1, 2, 3$), where λ_α are principal stretches, and the scalar function $f(\lambda_\alpha)$ satisfies the relations

$$f(1) = 0, \quad f'(1) = 1. \tag{2.61}$$

Members of this family of strain measures are:

Green–Lagrange strain tensor for $m = 2$: $E = \frac{1}{2}(U^2 - 1)$, (2.62)

Almansi strain tensor for $m = -2$: $e = \frac{1}{2}(1 - U^{-2})$, (2.63)

Biot strain tensor for $m = 1$: $U - 1$, (2.64)

Hencky strain tensor: $H = \ln U$. (2.65)

The tensors expressed by the deformation gradient F or Lagrangian and Eulerian triads

$$C = F^T F = U^2 = \sum_{i=1}^{3} \lambda_\alpha^2 N^\alpha \otimes N^\alpha, \tag{2.66}$$

and

$$B = FF^T = V^2 = \sum_{i=1}^{3} \lambda_\alpha^2 n^\alpha \otimes n^\alpha, \tag{2.67}$$

are known as the right and left Cauchy–Green tensors, respectively; where $\otimes$ signifies the Kronecker product. Eulerian strain tensors can be expressed in terms of the left Cauchy–Green tensor B as

$$e = \frac{1}{2}(g - B^{-1}). \tag{2.68}$$

The deformation rate tensor d and the spin tensor ω can be expressed as functions of the deformation gradient F as follows:

$$d = \{\dot{F}F^{-1}\}_s, \tag{2.69}$$

$$\omega = \{\dot{F}F^{-1}\}_A, \tag{2.70}$$

where $\{\cdot\}_s$ denotes the symmetric part and $\{\cdot\}_A$ the skew-symmetric part.

2.3 Stress Measures

The stress tensor σ_{ij} defined previously, determines the state of stress at the point x inside the deformed element of the body. The tensor σ_{ij} is called the Cauchy stress tensor and is determined by the relation

$$dP_i = \sigma_{ij} n_j ds. \tag{2.71}$$

The vector dP denotes here the force acting on the deformed element of the surface area ds and external normal n.

In order to build suitable stress tensors associated with the initial configuration we consider the element at the deformed state ds and undeformed state dS. Piola (1833) and Kirchhoff (1852) have defined two stress tensors. The first Piola–Kirchhoff tensor (nonsymmetric) one obtains by representing the actual force dP, acting on the deformed element ds to initial element dS, by the relation

$$dP_i = T_{iJ} N_J dS. \tag{2.72}$$

The second Piola–Kirchhoff tensor (symmetric) one obtains by replacing the actual force dP by the vector $d\tilde{P}$. The vector $d\tilde{P}$ is related to dP in the same way that the material vector dX of particle X is related to the post-deformation vector dx at point x

$$dP_I = \frac{\partial X_I}{\partial X_\alpha} dP_\alpha = S_{IJ} N_J dS. \tag{2.73}$$

The first Piola–Kirchhoff stress, also termed the nominal stress tensor, is inconvenient to use in constitutive equations, because it is nonsymmetric while the corresponding strain tensor is always symmetric.

Finally we can take as basic stress tensors: the Cauchy stress tensor in spatial description and the second Piola–Kirchhoff stress in material description. They are connected by the relation

$$S_{IJ} = \frac{\rho_0}{\rho} \frac{\partial X_J}{\partial x_j} \frac{\partial X_I}{\partial x_i} \sigma_{ij} \quad \text{or} \quad \sigma_{ij} = \frac{\rho}{\rho_0} \frac{\partial x_i}{\partial X_J} \frac{\partial x_i}{\partial X_J} S_{IJ} \tag{2.74}$$

where ρ_0 and ρ are reference and current mass densities, respectively. The relation (2.74) in curvilinear coordinate system using the definition (2.5) is expressed as

$$S = J\sigma. \tag{2.75}$$

Using the Kirchhoff tensor the equilibrium condition in the undeformed configuration (Lagrange description) has the different form

$$\frac{\partial}{\partial X_K} \left[S_{KM} \frac{\partial x_i}{\partial X_M} \right] + F_{oi} = 0, \tag{2.76}$$

and using the relation $x_i = X_I + u_i$

$$\frac{\partial}{\partial X_K} \left[S_{KM} \left(\delta_{IM} + \frac{\partial u_i}{\partial X_M} \right) \right] + F_{oi} = 0. \tag{2.77}$$

By F_{oi} are denoted the body force components in the system x_i per unit of initial volume.

The rate of stress work per unit mass

$$\dot{w} = \frac{1}{\rho_0} \text{tr}\{\boldsymbol{S}\dot{\boldsymbol{E}}\} = \frac{1}{\rho} \text{tr}\{\boldsymbol{\sigma}\boldsymbol{d}\}, \tag{2.78}$$

which is invariant under change of strain measure and reference configuration, could be used to generate in an arbitrary curvilinear coordinate system stress measures conjugate to the family of strain measures defined in (2.58). In Eq. (2.78) tr$\{\cdot\}$ means the trace operation. If with $\boldsymbol{\Sigma}$ we denote the general stress in the Lagrangian triad

$$\boldsymbol{\Sigma} = \sum_{\alpha,\beta=1}^{3} \Sigma_{\alpha\beta} \boldsymbol{N}^\alpha \otimes \boldsymbol{N}^\beta, \tag{2.79}$$

and write the Kirchhoff stress in the Eulerian triad

$$\boldsymbol{S} = \sum_{\alpha,\beta=1}^{3} S_{\alpha\beta} \boldsymbol{n}^\alpha \otimes \boldsymbol{n}^\beta \tag{2.80}$$

then using the identity

$$\frac{1}{\rho_0} \text{tr}\{\boldsymbol{S}\dot{\boldsymbol{E}}\} = \frac{1}{\rho_0} \text{tr}\{\boldsymbol{\Sigma}\dot{\tilde{\boldsymbol{E}}}\}, \tag{2.81}$$

it can be shown that the stress $\boldsymbol{\Sigma}$ conjugate to the general strain measure $\tilde{\boldsymbol{E}}$ may be written (in component form)

$$\begin{aligned}\Sigma_{\alpha\alpha} &= \frac{S_{\alpha\alpha}}{\lambda_\alpha f'(\lambda_\alpha)}, \\ \Sigma_{\alpha\beta} &= \frac{\lambda_\beta^2 - \lambda_\alpha^2}{f(\lambda_\beta) - f(\lambda_\alpha)} \frac{S_{\alpha\beta}}{2\lambda_\alpha \lambda_\beta}, \quad \alpha \neq \beta, \quad \lambda_\alpha \neq \lambda_\beta\end{aligned} \tag{2.82}$$

where no summation over repeated indices is implied.

For example, the stress conjugate to the logarithmic strain $\boldsymbol{H}$ can be expressed as

$$\begin{aligned}\Sigma_{\alpha\alpha}^L &= S_{\alpha\alpha}, \\ \Sigma_{\alpha\beta}^L &= \frac{\lambda_\alpha^2 - \lambda_\beta^2}{\ln(\lambda_\alpha/\lambda_\beta)} \frac{S_{\alpha\beta}}{2\lambda_\alpha \lambda_\beta}, \quad \alpha \neq \beta, \quad {}^\bullet\lambda_\alpha \neq \lambda_\beta\end{aligned} \tag{2.83}$$

(no summation indices).

For an isotropic elastic material the principal axes of the Kirchhoff stress tensor coincide with the Eulerian triad and equation (2.83) takes the simple form

$$\Sigma_{\alpha\beta}^L = S_{\alpha\beta} \tag{2.84}$$

which in direct notation can be written

$$\boldsymbol{\Sigma}^L = \boldsymbol{R}^T \boldsymbol{S} \boldsymbol{R}. \tag{2.85}$$

2.4 Final Remarks

In formulation of constitutive equations one should use conjugate variables, i.e. the pairs of stress tensor τ and respective deformation rate tensor de. For each configuration (along the trajectory) the infinitesimal increase of the work dw caused by deformation of the material element is expressed by the scalar product $dw = \tau de$. The conjugate variable pairs most often used are given in spatial formulation by Cauchy stress tensor and deformation rate tensor $(\sigma_{ij}-d_{ij})$; and in material formulation by Kirchhoff stress tensor and Green–Lagrange strain rate tensor $(S_{KM}-\dot{E}_{KM})$. The rate of energy dissipated in the volume v in the deformed configuration and in the volume V occupied by the same material in initial configuration is then the same and is equal to

$$\int_v \sigma_{ij} d_{ij} dv = \int_V S_{KM} \dot{E}_{KM} dV. \tag{2.86}$$

3 Temperature

The main purpose of the solution of heat exchange problems is to determine the amount of heat Q transferred in the system analysed. Such a system is usually bounded by a certain surface S and in practical cases one calculates an amount of heat transferred through that surface. Besides the total amount of heat Q one often uses the concept of heat flux. By the heat flux will be meant the magnitude

$$q = \lim_{\Delta S \to 0} \frac{\Delta Q_h}{\Delta S}, \qquad (3.1)$$

where ΔQ_h denotes an amount of heat transferred through the surface element ΔS and is referred to unit time. If heat exchange is stationary, then it does not depend explicitly on time. In the case when q is also independent of the position of the element ΔS on the surface then it is a constant and is given by

$$q = \frac{Q_h}{S}. \qquad (3.2)$$

The total amount of heat exchanged through the surface S in time t can then be calculated by the formula

$$Q = Q_h t = qSt.$$

In general Q is given by

$$Q = \int_S qt\, dS,$$

which reduces to (3.2) if q is a constant.

3.1 Heat Conduction

Heat conduction is governed by the Fourier law, which says that the heat flux vector is parallel to the temperature gradient. In one-dimensional situations this law can be expressed as

$$q = -k\frac{dT}{dx}, \qquad k > 0. \qquad (3.3)$$

The minus sign in Eq. (3.3) indicates that the heat always flows from hot to cold and not vice-versa. The coefficient of proportionality k is called the heat conduction coefficient and is a scalar-valued function which depends upon the properties of the material. If the heat transfer direction is normal to the surface S, then the amount of heat conducted in unit time through this surface is

$$Q_h = -kS\frac{dT}{dx}. \tag{3.4}$$

3.2 Heat Convection

A frequently encountered case of practical significance is the heat exchange between a solid wall and an adjacent gas or liquid. Heat exchange in fluids takes place by convection, but near the wall there exists a very thin layer in which heat exchange takes place by conduction. This fact is best illustrated by Fig. 3.1 which presents the temperature distribution in a fluid near a heated wall. In the case where heat exchange is stationary, the heat is transferred from the wall towards the centre of the fluid. If the intensity of heat transferred is higher, then the drop in temperature per unit length in a direction perpendicular to the wall is lower. Near the wall one observes a significant drop of temperature because in the thin boundary layer conduction plays a decisive role and heat exchange is less intensive in the boundary layer than in areas remote from the wall, where convection also takes place.

The phenomenon described above is known as heat convection. Mathematically it is described by the Newton equation

$$q = \alpha(T_w - T_f) \tag{3.5}$$

where T_w is the wall temperature, T_f is the fluid temperature at a sufficiently great distance from the wall; and the method of determination of T_f is usually precisely laid down. The magnitude α determining the heat exchange intensity is called the heat convection coefficient.

If the heat is convected to the wall from the fluid, then $T_w < T_f$ and in Eq. (3.5) $T_f - T_w$ is substituted for $T_w - T_f$. Although the Newton equation has a simple form, the experimental determination of the magnitude of the α coefficient is very difficult, because convection phenomena are complex.

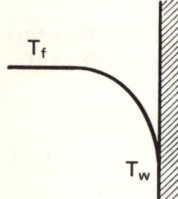

Figure 3.1. Temperature distribution in a fluid near a wall.

3.3 Heat Radiation

The concept of a black body is essential to radiation theory. A black body is a hypothetical body which absorbs all radiant energy falling on it, transmitting and reflecting nothing. Heat radiation occurs in accordance with the Stefan–Boltzmann law which says that the energy radiated by the black body is proportional to the fourth power of the absolute temperature of that body.

Mathematically this law is expressed by the formula

$$q = C_0 \left[\frac{T^a}{100}\right]^4 \tag{3.6}$$

where C_0 is the so-called radiation coefficient of the black body and T^a is the absolute temperature. The heat radiated through the surface S per unit time is

$$Q_h = C_0 S \left[\frac{T^a}{100}\right]^4. \tag{3.7}$$

Real bodies are not black bodies and at a given temperature will radiate less energy than a black body. If the ratio of the energy radiated by the real body to the energy radiated by the black body in the same conditions does not depend on radiation wavelength, then this body is called a grey body. The heat exchange between grey bodies is described by the equation

$$Q_{1-2} = C_0 S_1 \phi_{1-2} \left[\left(\frac{T_1^a}{100}\right)^4 - \left(\frac{T_2^a}{100}\right)^4\right], \tag{3.8}$$

where T_1^a and T_2^a are the absolute temperatures of the bodies radiating the heat, S_1 is the surface of the body at temperature T_1, and ϕ_{1-2} is the coefficient taking into consideration the deviation of the properties of the analysed body from the properties of the black body and the geometrical system of the two bodies.

Heat exchange based on pure conduction, convection or radiation very rarely holds in practice. These three fundamental kinds of heat exchange normally appear in various combinations. A common case is the exchange of heat through a solid wall by a combination of radiation and convection. In this case one introduces a substitute coefficient of heat exchange by radiation α_r, which is defined as follows

$$\alpha_r = \frac{Q_{1-2}}{S_1(T_1 - T_0)} \tag{3.9}$$

where Q_{1-2} is the heat exchanged by radiation, given by Eq. (3.8), T_1 is the wall temperature, and T_0 is the reference temperature. The reference temperature T_0 does not have to be equal to the temperature T_2 appearing in Eq. (3.8), and in the case of convection and radiation one puts it equal to the fluid temperature T_f. Eq. (3.9) can then be rewritten in the form

$$\alpha_r = \frac{C_0 \phi_{1-2}\left[\left(\frac{T_1^a}{100}\right)^4 - \left(\frac{T_2^a}{100}\right)^4\right]}{T_1 - T_f}. \tag{3.10}$$

Heat exchange by both convection and radiation can be described by the relation

$$q = (\alpha + \alpha_r)(T_w - T_f) \qquad (3.11)$$

where α_r is the heat radiation coefficient described by Eq. (3.9) or (3.10), and T_w and T_f are wall and fluid temperatures respectively.

3.4 Temperature Field in a Heat-Conducting Body

A heat conduction problem is solved if temperature distribution in the body is known, that is if the function

$$T = f(x, y, z, t) \qquad (3.12)$$

is determined at all points of the considered body. If the temperature field is dependent on time, then the temperature field is called nonstationary and heat conduction is also nonstationary. Otherwise the temperature field is called stationary and heat conduction is steady.

Temperature is a scalar-valued function and for its determination a number, being a measure of this magnitude, suffices. A set of points of uniform temperature is called an isothermal surface. Isothermal surfaces for different temperatures cannot intersect each other. The direction of temperature gradient is the same as the direction of the normal to the isothermal surface. A measure of temperature gradient is the ratio of temperature increment along the normal to an isotherm to the segment of this normal

$$\operatorname{grad} T = \frac{\partial T}{\partial n}.$$

Temperature gradient is connected with the amount of heat transferred by conduction in agreement with the Fourier law

$$\boldsymbol{q} = -k \operatorname{grad} T. \qquad (3.13)$$

Eq. (3.13) defines the heat flux vector. The heat flux vector is, as is temperature gradient, a vector-valued function.

In Cartesian coordinates x, y, z vector $\boldsymbol{q}$ has three components q_x, q_y and q_z

$$\begin{aligned} q_x &= -k \frac{\partial T}{\partial x}, \\ q_y &= -k \frac{\partial T}{\partial y}, \\ q_z &= -k \frac{\partial T}{\partial z}. \end{aligned} \qquad (3.14)$$

For nonstationary heat conduction the magnitude of $\boldsymbol{q}$ changes with time. The thermal balance in the solid element considering only heat fluxes leads to the expression*

$$\frac{\partial}{\partial x}\left(k\frac{\partial T}{\partial x}\right)+\frac{\partial}{\partial y}\left(k\frac{\partial T}{\partial y}\right)+\frac{\partial}{\partial z}\left(k\frac{\partial T}{\partial z}\right)+q_v=c_p\rho\frac{\partial T}{\partial t}, \qquad (3.15)$$

where q_v is the heat source

$$q_v = \lim_{\Delta V \to \infty} \frac{\Delta Q_h}{\Delta V}, \qquad (3.16)$$

and ΔQ_h is the heat generated in unit time in the volume ΔV of the system considered, ρ is the density of the medium and c_p is a material constant called the specific heat at constant pressure. c_p is a measure of heat generated in unit volume of the body during changes of temperature under conditions of constant pressure. Eq. (3.15) is the heat conduction equation in an isotropic medium taking into consideration internal heat generation. Eqs. (3.13) and (3.15) for an anisotropic case can be expressed in the forms (3.17a and b) respectively

$$q_i = -k_{ij}T_{,j}, \qquad (3.17a)$$

$$k_{ij}T_{,ij} + q_v = c_p\rho\dot{T}, \qquad (3.17b)$$

where k_{ij} are the components of the heat conductivity tensor which characterizes the material.

If heat conductivity k is a function of temperature, then Eq. (3.15) is nonlinear. Assuming $k = $ const, Eq. (3.15) can be written in the form

$$\frac{k}{c_p\rho}\nabla^2 T + \frac{q_v}{c_p\rho} = \frac{\partial T}{\partial t}, \qquad (3.18)$$

where

$$\nabla^2 T = \frac{\partial^2 T}{\partial x^2} + \frac{\partial^2 T}{\partial y^2} + \frac{\partial^2 T}{\partial z^2},$$

is the Laplace operator.

The expression

$$\frac{k}{c_p\rho} = a$$

is called the coefficient of thermal diffusion or temperature conduction.

For many practical problems it is convenient to use a system of coordinates other than Cartesian. The heat conduction equation in cylindrical coordinates r, θ, z, has the form

$$a\left(\frac{\partial^2 T}{\partial r^2}+\frac{1}{r}\frac{\partial T}{\partial r}+\frac{1}{r^2}\frac{\partial^2 T}{\partial \theta^2}+\frac{\partial^2 T}{\partial z^2}\right)+\frac{q_v}{c_p\rho}=\frac{\partial T}{\partial t}, \qquad (3.19)$$

---

* The total energetic balance can be found in Part IV.

and in spherical coordinates r, θ, φ

$$\frac{1}{r}\frac{\partial^2(rT)}{\partial r^2} + \frac{1}{r^2 \sin\theta}\frac{\partial}{\partial \theta}\left(\sin\theta \frac{\partial T}{\partial \theta}\right) + \frac{1}{r^2 \sin^2\theta}\frac{\partial^2 T}{\partial \varphi^2} + \frac{q_v}{k} = \frac{1}{a}\frac{\partial T}{\partial t}.$$

In order to solve the heat conduction equation it is necessary to know the initial and boundary conditions. The initial condition prescribes the temperature at time $t = 0$, that is

$$T(x, y, z, 0) = f(x, y, z). \tag{3.20}$$

The boundary conditions describe the heat exchange at the boundary of the body and are given by one of three possible cases.

1. The temperature distribution on the boundary of the body at any time t

$$T_w(t) = g(t) \tag{3.21}$$

where $T_w(t)$ means the body surface temperature. This condition is called a boundary condition of the first kind.

2. The heat flux is determined at each point of the body surface

$$q_w(t) = h(t). \tag{3.22}$$

This condition is called a boundary condition of the second kind.

3. The temperature of the surrounding medium and the relation describing the heat exchange between the heat-conducting body and the surroundings are known. The heat exchange between the heat-conducting body and its surroundings takes place by convection, radiation or by both of these phenomena and is most conveniently described by the Newton equation (3.5).

The Newton equation written for the surface element dS

$$dQ_h = \alpha(T_w - T_f)\,dS \tag{3.23}$$

shows the amount of heat exchanged by the element with the surroundings. On the other hand the same amount of heat has to be conducted at the boundary of the body, i.e.

$$dQ_h = -k(\operatorname{grad} T)_w\,dS \tag{3.24}$$

where $(\operatorname{grad} T)_w$ denotes the magnitude of the temperature gradient between the boundary of the body and the surroundings.

Comparison of the above two expressions for dQ_h gives

$$(\operatorname{grad} T)_w = -\frac{\alpha}{k}(T_w - T_f). \tag{3.25}$$

The above condition is called the boundary condition of the third kind. In the mathematical theory of heat conduction the ratio α/k is often denoted by h and called the heat exchange coefficient.

Some authors also introduce a fourth kind of boundary condition covering heat exchange with surroundings by conduction. This is a particular case of (3) above. In heat exchange with surroundings by conduction the body surface temperature T'_w and the surroundings temperature T''_w are identical

$$T'_w(t) = T''_w(t). \tag{3.26}$$

Moreover magnitudes of heat fluxes on the surface separating the body considered and the surroundings are identical

$$-k'\left(\frac{\partial T}{\partial n}\right)'_w = -k''\left(\frac{\partial T}{\partial n}\right)''_w. \tag{3.27}$$

3.5 Navier–Stokes Equation

The Navier–Stokes equation expresses combined convective and conductive transfer of thermal energy in regions containing a moving fluid. This problem is usually connected with nonlinear deformation processes, for example deformations in welding and casting. The transport of thermal energy in the fluid is described by

$$\rho c_p\left(\frac{\partial T}{\partial t} + v_j\frac{\partial T}{\partial x_j}\right) = \frac{\partial}{\partial x_j}\left(k_{ij}\frac{\partial T}{\partial x_j}\right) + q_v \tag{3.28}$$

where v_j is the fluid velocity.

Density changes are allowed to occur in the fluid in response to changes in the temperature according to the relation

$$\rho = \rho_0[(1 - \beta(T - T_0)] \tag{3.29}$$

where β is the coefficient of volumetric thermal expansion, and subscript 0 refers to the reference conditions. For the incompressible case the conservation of mass equation is written in the form

$$\frac{\partial v_i}{\partial x_i} = 0. \tag{3.30}$$

If g is the acceleration due to gravity, then the balance of linear momentum may be expressed as

$$\rho\left(\frac{\partial v_i}{\partial t} + v_j\frac{\partial v_i}{\partial x_j}\right) = \frac{\partial \tau_{ij}}{\partial x_j} - \rho g \beta(T - T_0), \tag{3.31}$$

where

$$\tau_{ij} = p\delta_{ij} + \mu\left(\frac{\partial v_i}{\partial x_j} + \frac{\partial v_j}{\partial x_i}\right), \tag{3.32}$$

and p is the pressure, and μ is the viscosity.

To complete the formulation of the boundary-value problem, suitable boundary conditions for the dependent variables are required. For the hydrodynamic part of the problem either velocity components or the total surface stress (or traction) must be specified on the boundary of the fluid region. The thermal part of the problem requires a temperature or heat flux to be specified on all parts of the boundary. Symbolically, these conditions are expressed by

$$v_i = f_i(s) \quad \text{on } S_v, \tag{3.33}$$

$$t_i = \tau_{ij}(s)n_j(s) \quad \text{on } S_t, \quad S = S_v \cup S_t. \tag{3.34}$$

In Eqs. (3.33) and (3.34) s denotes a generic point on the boundary, and n_j are the components of the outward unit normal to the boundary.

4 Thermodynamical Considerations

4.1 Thermomechanical Process

Suppose that the thermodynamical state of each material element is uniquely defined by the values of a finite set of state variables even in an irreversible process. Such a phenomenological theory is, of course, restricted to a limited class of materials on the one hand and to processes running not too far from thermodynamical equilibrium on the other hand. A thermomechanical process starts in the initial state B_{t_0} of the body which is characterized by the initial configuration and by the initial thermodynamical state of each material element. The initial state is described by the prescribed thermomechanical boundary conditions, the prescribed body forces and energy sources acting inside and on the surface of the body. The process is governed by the field equations (balance equations) and by the constitutive law of the material. We focus our attention on the constitutive law which governs the local thermomechanical process within the thermodynamical state space.

Concerning these local thermomechanical processes we can distinguish on the first level between strictly reversible processes governed uniquely by thermodynamical state equations and other processes. From the phenomenological point of view we can subdivide the second class into four subclasses (Lehmann, 1979, 1980, 1982a,b, 1984):

1. Plastic deformations characterized by constrained equilibrium states
2. Internal processes leading to changes of the internal structure of the material
3. Thermally activated processes (unconstrained equilibrium states) leading to unlimited creep processes (high temperature creep or long term creep)
4. Viscous (damping) processes.

The internal process 2 may be coupled with processes of the kind 1 or 3. However, it can also occur independently; for instance, solid phase transfomation, recrystallization, or recovery. It may or may not be connected with deformations. Finally, damping processes may be correlated with all other kinds of processes including reversible processes as, for instance, in viscoelastic deformations. These considerations suggest a material model as shown in Fig. 4.1.

The particular structure and the mutual arrangement of the different elements is determined by the respective constitutive laws. Some particular cases of such constitutive laws will be discussed later. A real thermomechanical process

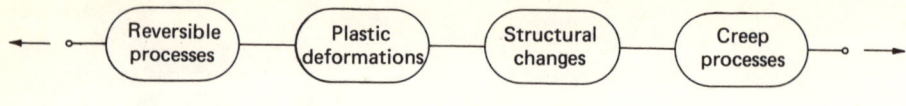

Damping parallel arranged

Figure 4.1. Model of the material (Lehmann 1984).

carries the body from the initial state B_{t_0} into the current state B_t. We attach to the current state B_t of the body an accompanying fictitious reference state B_t^* by means of a fictitious reversible process which carries each material element from its actual thermodynamical state into an unstressed state at reference temperature T_0 (Fig. 4.2). During this fictitious process the internal variables are kept constant in order to ensure a unique definition of reversible energy.

There is no real or fictitious process leading from the initial state B_{t_0} to the reference state B_t^*. Therefore it becomes unnecessary to introduce any strain tensor defining the non-reversible deformations uniquely. We need, however, a unique decomposition of the rate of mechanical work into its reversible part and its remaining parts. This means at the same time a unique decomposition of the deformation rate into corresponding parts. Furthermore we require a definite measure for the reversible strain serving as the thermodynamical state variable.

In the following we shall first discuss the mechanical and thermodynamical framework for the formulation of the constitutive law in our sense of a phenomenological theory. Finally we shall point to certain coupling effects occurring in some thermomechanical processes.

4.2 Formulation of the Constitutive Law

A thermomechanical process in a body can be discussed with respect to spatial description or material description. Using a spatial description a material point keeps its coordinate x_i during the whole process. The base vectors and the metric in the initial configuration of the body are denoted by G_i and G_{ik}, respectively. The corresponding quantities in the current configuration are g_i and g_{ik}. The deformation of the body can be measured by the quantities

$$q_{ik} = G_{ir} g_{rk}, \qquad (q^{-1})_{ik} = g_{ir} G_{rk}. \tag{4.1}$$

Relating q_{ik} to the current configuration, we obtain the tensor

$$\boldsymbol{q} = q_{ik} \boldsymbol{g}_i \otimes \boldsymbol{g}_k. \tag{4.2}$$

Furthermore the deformation rate is expressible in the form

$$\boldsymbol{d} = \tfrac{1}{2}(q^{-1})_{ir}(\dot{q})_{r.k} \boldsymbol{g}_i \otimes \boldsymbol{g}_k = d_{ik} \boldsymbol{g}_i \otimes \boldsymbol{g}_k. \tag{4.3}$$

$(\dot{\ }) = \dfrac{\partial}{\partial t}$ denotes the material derivative with respect to time t which differs from the substantial time derivative in the spatial description. It corresponds to one of the Oldroyd derivations (Oldroyd 1950; Lehmann 1960, 1962). This material

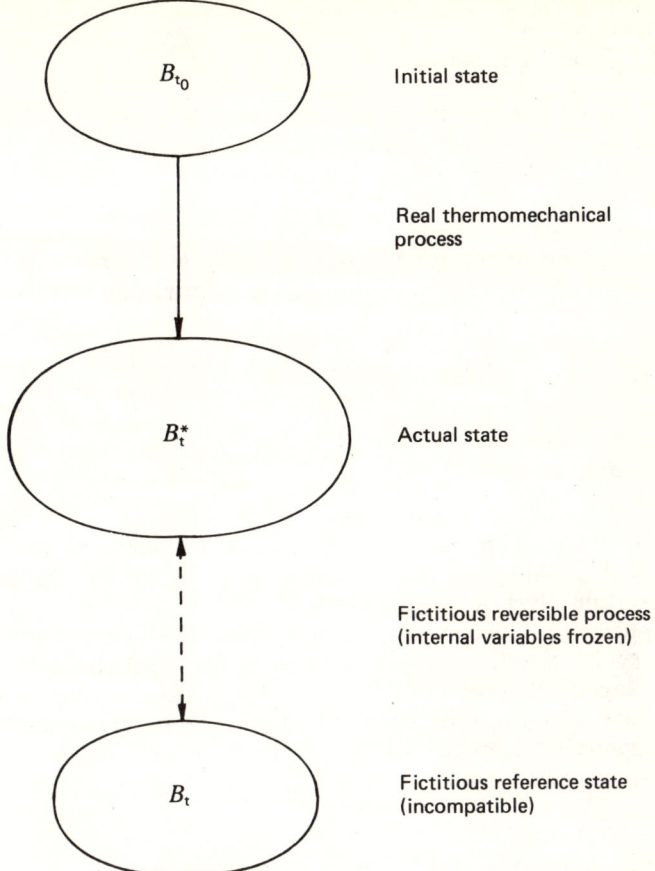

Figure 4.2. Thermomechanical process.

time derivative is objective, in contradiction to the substantial time derivative, since rigid body rotations are eliminated from the very beginning by the introduction of a material coordinate system.

Using the definition (4.3) for the deformation rate and the definition of the Cauchy stress tensor in the current configuration with respect to the spatial coordinate system

$$\boldsymbol{\sigma} = \sigma_{ik} \boldsymbol{g}_i \otimes \boldsymbol{g}_k, \tag{4.4}$$

the specific rate of work can be written as

$$\dot{w} = \frac{1}{\rho} \operatorname{tr}\{\boldsymbol{\sigma d}\} = \frac{1}{\rho} \sigma_{ir} (q^{-1})_{rk} \frac{1}{2} (\dot{q})_{k.i}. \tag{4.5}$$

This means that

$$\sigma_{i.k} = \sigma_{ir}(q^{-1})_{rk} \quad \text{and} \quad \tfrac{1}{2}q_{ik} \quad \text{or} \quad \tfrac{1}{2}(q_{ki} - \delta_{ki}) \tag{4.6}$$

can be considered as a conjugate pair of stress and strain.

The total work rate can be decomposed into the rate of reversible work $\dot{w}^{(r)}$ and the rate of remaining work $\dot{w}^{(i)}$ according to

$$\dot{w} = \dot{w}^{(r)} + \dot{w}^{(i)}. \tag{4.7}$$

If the reversible processes are without damping we obtain from (4.7) also a corresponding additive decomposition of the deformation rate. In this case

$$\frac{1}{\rho}\operatorname{tr}\{\boldsymbol{\sigma d}\} = \dot{w} = \dot{w}^{(r)} + \dot{w}^{(i)} = \frac{1}{\rho}\operatorname{tr}\{\boldsymbol{\sigma d}^{(r)}\} + \frac{1}{\rho}\operatorname{tr}\{\boldsymbol{\sigma d}^{(i)}\}. \tag{4.8}$$

This means

$$\boldsymbol{d} = \boldsymbol{d}^{(r)} + \boldsymbol{d}^{(i)}. \tag{4.9}$$

For thermodynamical reasons the rate of reversible work must also be expressible as the scalar product of a suitably defined stress tensor and the time derivative of the conjugated strain tensor. At the same time this strain tensor must fulfil certain physical requirements in order to define reversible deformations independent of accompanying non-reversible deformations. In many papers this problem is discussed from a different point of view (see for instance, Green and Naghdi 1965, 1971). Concerning the possible decompositions of total strain and total strain rate we shall only discuss two approaches.

One frequently used procedure starts with the multiplicative decomposition of the deformation gradient

$$\boldsymbol{F} = \boldsymbol{F}^{(r)}\boldsymbol{F}^{(i)}. \tag{4.10}$$

The polar decomposition of $\boldsymbol{F}^{(r)}$ (see Eq. (2.6)) leads to

$$\boldsymbol{F}^{(r)} = \boldsymbol{V}^{(r)}\boldsymbol{R}^{(r)}. \tag{4.11}$$

From (4.9) we derive an additive decomposition of the total strain rate

$$\boldsymbol{d} = \{\dot{\boldsymbol{F}}\boldsymbol{F}^{-1}\}_s = \{\dot{\boldsymbol{F}}^{(r)}\boldsymbol{F}^{-1(r)}\}_s + \{\boldsymbol{F}^{(r)}\dot{\boldsymbol{F}}^{(i)}\boldsymbol{F}^{-1(i)}\boldsymbol{F}^{-1(r)}\}_s. \tag{4.12}$$

Using the polar decomposition, the first term of the right hand side reads

$$\begin{aligned}\{\dot{\boldsymbol{F}}^{(r)}\boldsymbol{F}^{-1(r)}\}_s &= \{\dot{\boldsymbol{V}}^{(r)}\boldsymbol{R}^{(r)}\boldsymbol{R}^{T(r)}\boldsymbol{V}^{-1(r)}\}_s + \{\boldsymbol{V}^{(r)}\dot{\boldsymbol{R}}^{(r)}\boldsymbol{R}^{T(r)}\boldsymbol{V}^{-1(r)}\}_s \\ &= \tfrac{1}{2}\{\dot{\boldsymbol{V}}^{(r)}\boldsymbol{V}^{-1(r)} + \boldsymbol{V}^{-1(r)}\dot{\boldsymbol{V}}^{(r)}\} + \tfrac{1}{2}\{\boldsymbol{V}^{(r)}\boldsymbol{\omega}^{(r)}\boldsymbol{V}^{-1(r)} \\ &\quad - \boldsymbol{V}^{-1(r)}\boldsymbol{\omega}^{(r)}\boldsymbol{V}^{(r)}\}.\end{aligned} \tag{4.13}$$

We see this expression depends on the arbitrary (local) spin of the reference configuration. The reason is that the spatial time derivative entering this expression is not objective. If, however, the reversible behaviour of the material is isotropic the expression for the work rate reduces to

$$\dot{w} = \frac{1}{\rho}\operatorname{tr}(\boldsymbol{\sigma}\boldsymbol{d}) = \frac{1}{\rho}\operatorname{tr}\left\{\boldsymbol{\sigma}\frac{1}{2}\{\dot{\boldsymbol{V}}^{(r)}\boldsymbol{V}^{-1(r)} + \boldsymbol{V}^{-1(r)}\dot{\boldsymbol{V}}^{(r)}\}\right\}$$
$$+ \frac{1}{\rho}\operatorname{tr}\left\{\boldsymbol{\sigma}\{\boldsymbol{F}^{(r)}\dot{\boldsymbol{F}}^{(i)}\boldsymbol{F}^{-1(i)}\boldsymbol{F}^{-1(r)}\}\right\}$$
$$= \frac{1}{\rho}\operatorname{tr}\left\{\{\boldsymbol{\sigma}\boldsymbol{V}^{-1(r)}\}\dot{\boldsymbol{V}}^{(r)}\right\} + \frac{1}{\rho}\operatorname{tr}\left\{\boldsymbol{\sigma}\{\boldsymbol{R}^{(r)}\{\dot{\boldsymbol{F}}^{(i)}\boldsymbol{F}^{-1(i)}\}_s\boldsymbol{R}^{T(r)}\}\right\}. \tag{4.14}$$

This means that the first term on the right hand side and therefore also the second term become independent of the reference configuration. In this case we can define

$$\boldsymbol{d}^{(r)} = \tfrac{1}{2}\{\dot{\boldsymbol{V}}^{(r)}\boldsymbol{V}^{-1(r)} + \boldsymbol{V}^{-1(r)}\dot{\boldsymbol{V}}^{(r)}\}, \tag{4.15}$$

as reversible strain rate and

$$\tilde{\boldsymbol{\sigma}} = \boldsymbol{\sigma}\boldsymbol{V}^{-1(r)} \quad \text{and} \quad \boldsymbol{V}^{(r)} \tag{4.16}$$

as the conjugate pair of stress and strain with respect to the reversible deformations. However, it must be emphasized once more that this is only possible in the case of isotropy of the reversible strains.

We avoid these restrictions when we base our considerations on a multiplicative decomposition of the tensor $\boldsymbol{q}$ writing

$$q_{jk} = q^{(i)}_{j.m}q^{(r)}_{mk}. \tag{4.17}$$

This leads again to an additive decomposition of the total strain rate according to

$$d_{jk} = \tfrac{1}{2}(q^{-1})_{jm}(\dot{q})_{m.k} = \tfrac{1}{2}(q^{-1(r)})_{jm}(\dot{q}^{(r)})_{m.k}$$
$$+ \tfrac{1}{2}(q^{-1})_{jm}(\dot{q}^{(r)})_{m.s}q^{(r)}_{sk} = d^{(r)}_{j.k} + d^{(i)}_{j.k}. \tag{4.18}$$

The partial strain rates $d^{(r)}_{j.k}$ and $d^{(i)}_{j.k}$ are in general nonsymmetric tensors. However, their sum is symmetric and only their symmetric parts enter the respective expressions for the partial work rates. Therefore we can write

$$\dot{w}^{(r)} = \frac{1}{\rho}\sigma_{ik}d^{(r)}_{ik} = \frac{1}{\rho}\sigma_{ir}(q^{-1(r)})_{rk}\tfrac{1}{2}(\dot{q}^{(r)})_{k.i}. \tag{4.19}$$

This means we can take

$$\tilde{\sigma}_{i.k} = \sigma_{ir}(q^{-1(r)})_{rk} \quad \text{and} \quad \tfrac{1}{2}q^{(r)}_{ik} \tag{4.20}$$

as the conjugate pair of stress and strain with respect to the reversible deformations even in the anisotropic case. In the case of isotropy, $\tilde{\sigma}_{i.k}$ becomes symmetric. Then we can also write

$$\dot{w}^{(r)} = \tilde{\sigma}_{ik}\tfrac{1}{2}\overset{\triangledown}{q}{}^{(r)}_{ki}, \tag{4.21}$$

where

$$\overset{\triangledown}{q}{}^{(r)}_{ki} = (\dot{q}^{(r)})_{k.i} + d^{(r)}_{kr}q_{ri} - d^{(r)}_{ri}q_{kr} = \{(\dot{q})_{k.i}\}_s,$$

represents the covariant time derivative which corresponds to the Zaremba–Jaumann time derivative in the space fixed coordinate system. In the isotropic case it also holds that

$$\dot{w}^{(r)} = \frac{1}{\rho} \sigma_{ik} \overset{\triangledown}{\varepsilon}_{ki}^{(r)} \tag{4.22}$$

with the logarithmic strain tensor

$$\varepsilon_{ik}^{(r)} = \tfrac{1}{2} (\ln q^{(r)})_{ik}. \tag{4.23}$$

For simplicity in the following we restrict ourselves to isotropy of the reversible processes using σ_{ik} and $\varepsilon_{ik}^{(r)}$ as a conjugated pair of stress and strain.

The first law of thermodynamics states

$$\dot{u} = \dot{w} - \frac{1}{\rho} q_{i,i} + r = \dot{w}^{(r)} + \dot{w}^{(i)} - \frac{1}{\rho} q_{i,i} + r \tag{4.24}$$

where u is the specific internal energy, q_i is the energy flux and r is the external heat supply. The energy flux includes heat flux and other energy fluxes which may occur, for instance due to diffusion of self-equilibrated microstress fields. These other energy fluxes are mostly small in solid bodies and therefore negligible in many cases. We shall neglect them hereafter.

Within the framework of the phenomenological theory presented here u must be expressible as a unique function of a finite set of thermodynamical state variables. This set may consist of reversible strain $\varepsilon_{ik}^{(r)}$, specific entropy S and a representative set of internal variables b, β_{ik}. Then we can write

$$u = u(\varepsilon_{ik}^{(r)}, S, b, \beta_{ik}). \tag{4.25}$$

Replacing $\varepsilon_{ik}^{(r)}$ and S by their conjugated state variables, i.e. by the stress σ_{ik} and the temperature T, by means of Legendre transformations we obtain the specific free enthalpy

$$\psi = u - \frac{1}{\rho_0} \sigma_{ik} \varepsilon_{ki}^{(r)} - TS = \psi(\sigma_{ik}, T, b, \beta_{ik}), \tag{4.26}$$

as a thermodynamic state function. From (4.26) we derive the thermic state equation

$$\varepsilon_{ik}^{(r)} = -\rho_0 \frac{\partial \psi}{\partial \sigma_{ki}} = \varepsilon_{ik}^{(r)}(\sigma_{ik}, T, b, \beta_{ik}) \tag{4.27}$$

and the caloric state equation

$$S = -\frac{\partial \psi}{\partial T} = S(\sigma_{ik}, T, b, \beta_{ik}). \tag{4.28}$$

Concerning the changes in specific free enthalpy we obtain from Eqs. (4.26) and (4.24) the expressions

$$\dot{\psi} = \dot{w}^{(r)} + \dot{w}^{(i)} - \frac{1}{\rho} q_{i,i} + r - \frac{1}{\rho_0} \overset{\triangledown}{\sigma}_{ik} \varepsilon_{ki}^{(r)} - \frac{1}{\rho_0} \sigma_{ik} \overset{\triangle}{\varepsilon}_{ki}^{(r)} - \dot{T}S - T\dot{S}$$

$$= \frac{\partial \psi}{\partial \sigma_{ik}} \overset{\triangledown}{\sigma}_{ik} + \frac{\partial \psi}{\partial T} \dot{T} + \frac{\partial \psi}{\partial b} \dot{b} + \frac{\partial \psi}{\partial \beta_{ik}} \overset{\triangledown}{\beta}_{ik}. \tag{4.29}$$

Finally from Eqs. (4.27), (4.28) and (4.29) we derive the balance equation for specific reversible work

$$\dot{w}^{(r)} = \frac{1}{\rho_0} \sigma_{ik} \overset{\triangledown}{\varepsilon}^{(r)}_{ki}$$
$$= -\sigma_{ik} \left\{ \frac{\partial^2 \psi}{\partial \sigma_{rs} \partial \sigma_{ik}} \overset{\triangledown}{\sigma}_{rs} + \frac{\partial^2 \psi}{\partial T \partial \sigma_{ik}} \dot{T} + \frac{\partial^2 \psi}{\partial b \partial \sigma_{ik}} \dot{b} + \frac{\partial^2 \psi}{\partial \beta_{rs} \partial \sigma_{ik}} \overset{\triangledown}{\beta}_{rs} \right\}, \quad (4.30)$$

the balance equation for remaining specific energy supply

$$\dot{w}^{(i)} - \frac{1}{\rho} q_{i,i} + r = -T \left\{ \frac{\partial^2 \psi}{\partial \sigma_{ik} \partial T} \overset{\triangledown}{\sigma}_{ik} + \frac{\partial^2 \psi}{\partial T^2} \dot{T} \right\} + \frac{\partial}{\partial b} \left\{ \psi - T \frac{\partial \psi}{\partial T} \right\} \dot{b}$$
$$+ \frac{\partial}{\partial \beta_{ik}} \left\{ \psi - T \frac{\partial \psi}{\partial T} \right\} \overset{\triangledown}{\beta}_{ik}, \quad (4.31)$$

and the balance equation for specific entropy (Gibbs equation)

$$T\dot{S} = \dot{w}^{(i)} - \frac{1}{\rho} q_{i,i} + r - \frac{\partial \psi}{\partial b} \dot{b} - \frac{\partial \psi}{\partial \beta_{ik}} \overset{\triangledown}{\beta}_{ik}. \quad (4.32)$$

In balance equation (4.32) we have to decompose the rate of the specific entropy into its reversible part $\dot{S}^{(r)}$ and its irreversible dissipative part $\dot{S}^{(d)}$ (entropy production)

$$T\dot{S} = T\dot{S}^{(r)} + T\dot{S}^{(d)}. \quad (4.33)$$

Concerning this decomposition within the frame of a phenomenological theory we have to distinguish four different classes of processes (Lehmann 1973, 1977, 1983a,b, 1984):

1. Strictly reversible, non-dissipative processes governed by state equations and representing a sequence of equilibrium states
2. Irreversible, dissipative processes characterized essentially by non-equilibrium states
3. Dissipative processes appearing as a sequence of equilibrium states
4. Non-dissipative processes appearing as a sequence of equilibrium states but not governed by state equations

At microscopic level only classes 1 and 2 occur, these can be treated within the frame of the classical theory of reversible or irreversible processes, respectively. The existence of class 3 is due to the fact that some irreversible processes at microscopic level may have very short relaxation times. These dissipative processes appear at the macroscopic level as a sequence of equilibrium states, for instance plastic deformation. The occurrence of processes of class 4 is a consequence of the fact that at the macroscopic level we are dealing with a so-called small state space. Therefore certain non-dissipative processes become dependent on the history of the process as, for instance, anisotropic hardening (or softening) due to inelastic deformations and in connection with storing and restoring of mechanical energy.

From these facts it follows that contributions to entropy production have to be defined within the constitutive law.
These contributions include:

1. The immediately dissipated specific work

$$\dot{w}^{(d)} = \dot{w}^{(i)} - \dot{w}^{(h)} \tag{4.34}$$

where $\dot{w}^{(h)}$ denotes the specific mechanical work stored in changes of the internal structure of the material.

2. The irreversible part of the heat flux

$$-\frac{1}{\rho T} q_i T_{,i}. \tag{4.35}$$

3. The entropy production $T\dot{\eta}$ due to other dissipative processes which may be involved in internal processes, in energy supply by sources, and (as far as they are not negligible) in energy fluxes different from heat flux. According to the second law of thermodynamics the entropy production cannot become negative. This means

$$T\dot{S}^{(d)} = \dot{w}^{(d)} - \frac{1}{\rho T} q_i T_i + T\dot{\eta} \geq 0. \tag{4.36}$$

These dissipative (rate-dependent and rate-independent) processes can be treated by means of so-called dissipative potentials. How this can be done will not be discussed here. We refer to Raniecki (1983).

Within the thermodynamical framework which is given by the relations (4.26) and (4.30)–(4.36) the constitutive law has to be defined. It consists of:

1. The state function for specific free enthalpy governing also the immediately reversible processes
2. Evolution laws for non-reversible deformations
3. Evolution laws for internal variables
4. Flux laws for energy (heat flux and possible other fluxes)
5. Laws of entropy production ($\dot{w}^{(d)}$, $T\dot{\eta}$)

Many different models of constitutive laws have been introduced in order to describe the inelastic behaviour of solid bodies, particularly of polycrystalline metals. Some of these are concerned with small deformations occurring in creep and relaxation processes. Others aim at large deformations in general processes. Another group deals with special problems connected with solid phase transformations occurring in quenching processes or in deformations of memory alloys (Miller, 1983; Delang et al., 1974). All these models fit the frame of the general material model given in Fig. 4.1. They emphasize various special features of this model. In this book we present only selected, frequently used, theories to demonstrate some basic points of view.

PART II

FUNDAMENTALS OF ELASTICITY AND PLASTICITY THEORY

5 Stress–Strain Curve

The uniaxial tension test is a convenient method for the determination of mechanical properties of a material. A typical curve in the system nominal stress–Cauchy strain is presented in Fig. 5.1, in which the following characteristic points are indicated: 1: proportional limit σ_p; 2: elastic limit σ_H; 3: yield point σ_0; 3–4: platform of ideal plasticity; 4–5: plastic hardening; 5: necking point; 6: rupture point. The character of stress–strain curves for various materials is different. For materials for which one does not observe a well-marked yield point, for example alloy steel, one introduces the notion of yield strength corresponding to some fixed elongation, for instance 0.2% ($\sigma_{0,2}$). Many materials do not give evidence of existence of a linear segment (annealed aluminium for example).

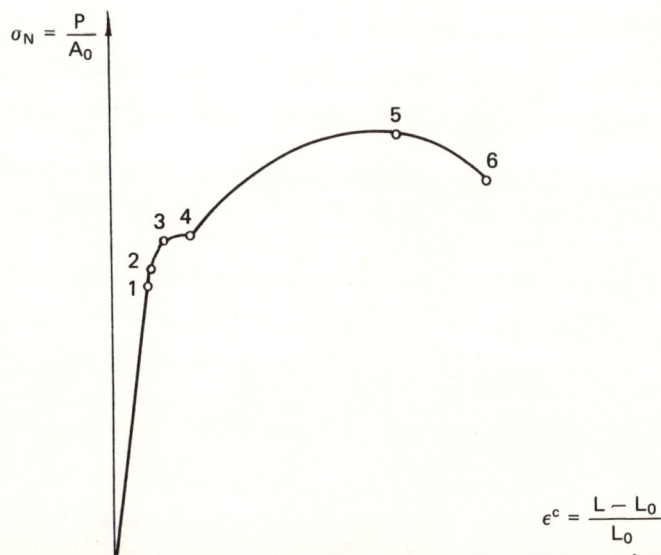

Figure 5.1. Stress–strain curve for soft steel.

A compression curve may be similarly obtained. One assumes that the proportional limit and yield point during compression are equal to the proportional limit and yield point in tension.

In general, all materials and their mechanical properties are sensitive to temperature and strain rate magnitude. For example, consider changes of yield point as a function of temperature and changes of σ–ε curve as a function of strain rate. Fig. 5.2 illustrates the influence of temperature on yield point for soft steel. Changes in the position of the yield point as a function of the strain rate for

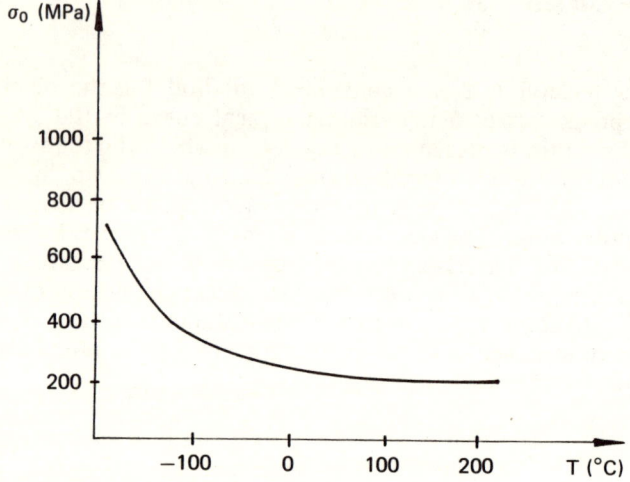

Figure 5.2. Yield stress as a function of temperature for soft steel (Maiden and Campbell 1958).

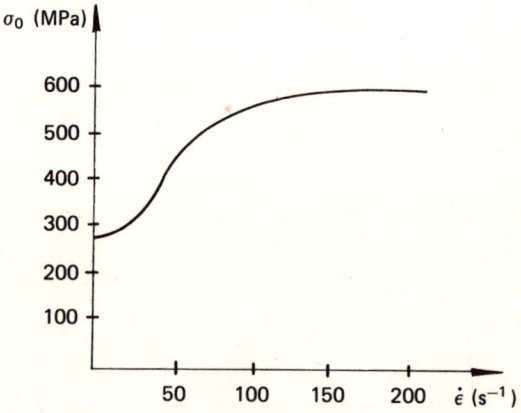

Figure 5.3. Influence of strain rate on yield point for 0.22% carbon steel (Clark and Duwez 1950).

0.22% carbon steel are presented in Fig. 5.3. Stress–strain–strain rate curves for aluminium are presented in Fig. 5.4. From these figures it is seen that changes of temperature and strain rate play an essential role in thermo-elasto-plastic analyses.

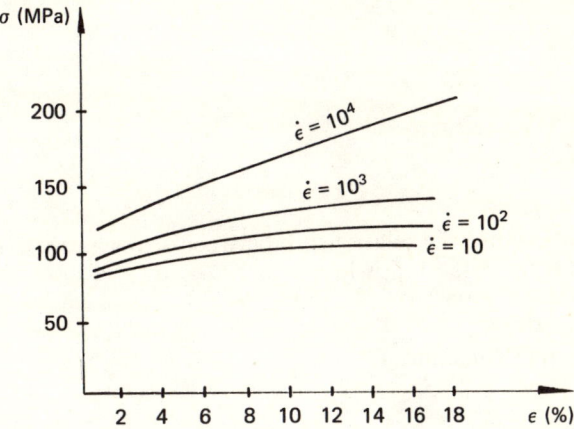

Figure 5.4. Stress–strain curve for aluminium for various strain rates (Hauser et al. 1960).

6 Elasticity

In the linear elastic range the relationship between stress and strain components for an isotropic solid is the following

$$\sigma_{ij} = \lambda \varepsilon_{kk} \sigma_{ij} + 2\mu \varepsilon_{ij}, \tag{6.1}$$

where λ is the Lame constant,

$$\lambda = \frac{E\nu}{(1 + \nu)(1 - 2\nu)} \tag{6.2}$$

$\mu = G$ is the shear modulus of elasticity, E is Young's modulus and ν is Poisson's ratio.

For anisotropic bodies, the stress–strain relation used in linear elastic stress analysis can be expressed as

$$\sigma_{ij} = C^e_{ijkl} \varepsilon_{kl}, \tag{6.3}$$

or

$$\varepsilon_{ij} = E_{ijkl} \sigma_{kl},$$

where C^e_{ijkl}, E_{ijkl} denote the material coefficients.

Nonlinear material behaviour in the elastic range is primarily due to the fact that the material properties, e.g. E, vary with temperature and strain rate $\dot{\varepsilon}$. Although the relationship between σ_{ij} and ε_{kl} is nonlinear in the global sense, linearity can still be considered to be valid for small strains.

For situations involving drastic temperature and strain rate changes the components of the total strain rate may be assumed to consist of the following components

$$\dot{\varepsilon}_{ij} = \dot{\varepsilon}^e_{ij} + \dot{\varepsilon}^T_{ij} + \dot{\varepsilon}^{e,T}_{ij} + \dot{\varepsilon}^{e,\dot{\varepsilon}}_{ij} \tag{6.4}$$

where $\dot{\varepsilon}^e_{ij}$ are components of the strain rate tensor due to elastic deformation, $\dot{\varepsilon}^{e,T}$ are components of the strain rate tensor due to temperature dependent material properties, $\dot{\varepsilon}^{e,\dot{\varepsilon}}_{ij}$ are components of the strain rate tensor due to strain rate dependent material properties, and $\dot{\varepsilon}^T_{ij}$ are components of the thermal strain rate tensors.

Functionally (6.4) can be expressed as

$$\varepsilon_{ij} = \varepsilon_{ij}(\sigma_{ij}, T, E_{ijkl}(T, \dot{\varepsilon})). \tag{6.5}$$

Applying the chain rule for partial differentation to (6.5) gives

$$\dot{\varepsilon}_{ij} = \frac{\partial \varepsilon_{ij}}{\partial \sigma_{ij}} \dot{\sigma}_{ij} + \frac{\partial \varepsilon_{ij}}{\partial T} \dot{T} + \frac{\partial \varepsilon_{ij}}{\partial E_{ijkl}} \left[\frac{\partial E_{ijkl}}{\partial T} \dot{T} + \frac{\partial E_{ijkl}}{\partial \dot{\bar{\varepsilon}}} (\dot{\bar{\varepsilon}})^{\cdot} \right]. \tag{6.6}$$

Since $\dot{\varepsilon}_{ij} = E_{ijkl}\dot{\sigma}_{kl}$ and $\dot{\varepsilon}^T_{ij} = \alpha_{ij}\dot{T}$ for thermal strain, where α_{ij} are linear thermal expansion tensor components, the above expression becomes

$$\dot{\varepsilon}_{ij} = E_{ijkl}\dot{\sigma}_{kl} + \alpha_{ij}\dot{T} + \left[\frac{\partial E_{ijkl}}{\partial T} \dot{T} + \frac{\partial E_{ijkl}}{\partial \dot{\bar{\varepsilon}}} (\dot{\bar{\varepsilon}})^{\cdot} \right] \sigma_{kl}. \tag{6.7}$$

The constitutive equation for elastic solids with nonlinear material properties can be derived as

$$\dot{\sigma}_{ij} = C^e_{ijkl}\dot{\varepsilon}_{kl} - C^e_{ijkl}\left(\alpha_{kl}\dot{T} + \left(\frac{\partial E_{klmn}}{\partial T} \dot{T} + \frac{\partial E_{klmn}}{\partial \dot{\bar{\varepsilon}}} (\dot{\bar{\varepsilon}})^{\cdot} \right) \sigma_{mn} \right). \tag{6.8}$$

7 Plasticity

7.1 Idealization of Tension Test

Linear idealizations of tension tests most often used in the literature are illustrated in Fig. 7.1. They show the following materials (a) rigid ideal-plastic, (b) ideal elasto-plastic, (c) linear hardening rigid-plastic, (d) linear hardening elasto-plastic. The respective equations have the form:

$$\sigma = \sigma_0, \tag{7.1a}$$

$$\sigma = \begin{cases} E\varepsilon & (\varepsilon \leq \sigma_0/E), \\ \sigma_0 & (\varepsilon \geq \sigma_0/E), \end{cases} \tag{7.1b}$$

$$\sigma = \sigma_0 + E_1 \varepsilon, \tag{7.1c}$$

$$\sigma = \begin{cases} E\varepsilon & (\varepsilon \leq \sigma_0/E), \\ E\varepsilon[1 - \omega(\varepsilon)] & (\varepsilon \geq \sigma_0/E). \end{cases} \tag{7.1d}$$

where $\omega(\varepsilon) = \dfrac{E - E_1}{E}\left(1 - \dfrac{\sigma_0}{E\varepsilon}\right)$.

Nonlinear idealizations of work-hardening materials are presented in Fig. 7.2. These are (a) work-hardening plastic material, (b) work-hardening rigid-plastic material, (c) work-hardening elasto-plastic material, (d) elasto-plastic Ramberg–Osgood material. The equations describing these idealizations are respectively

$$\sigma = k\varepsilon^n, \tag{7.2a}$$

$$\sigma = \sigma_0 + k\varepsilon^n, \tag{7.2b}$$

$$\sigma = \begin{cases} E\varepsilon & (\varepsilon \leq \sigma_0/E), \\ k\varepsilon^n & (\varepsilon \geq \sigma_0/E), \end{cases} \tag{7.2c}$$

$$\varepsilon = \frac{\sigma}{E} + k\left(\frac{\sigma}{E}\right)^m, \quad (m \geq 1). \tag{7.2d}$$

PLASTICITY 51

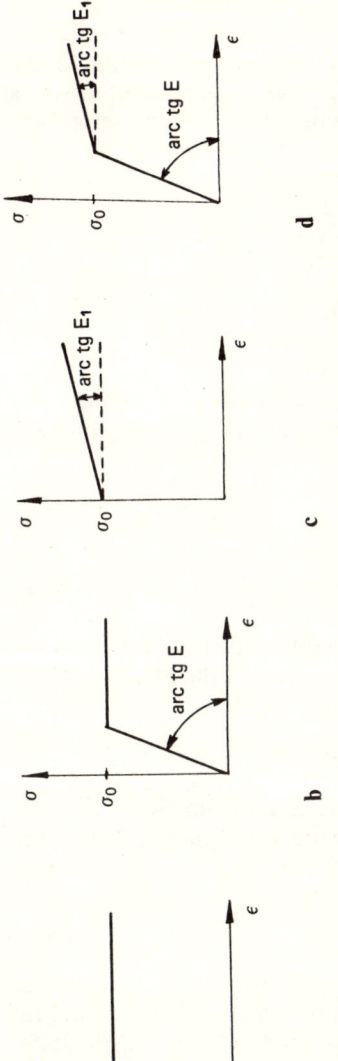

Figure 7.1a–d. Linear idealizations of tension tests. **a** Rigid ideal-plastic material. **b** Ideal elasto-plastic material. **c** Linear hardening rigid-plastic material. **d** Linear hardening elasto-plastic material.

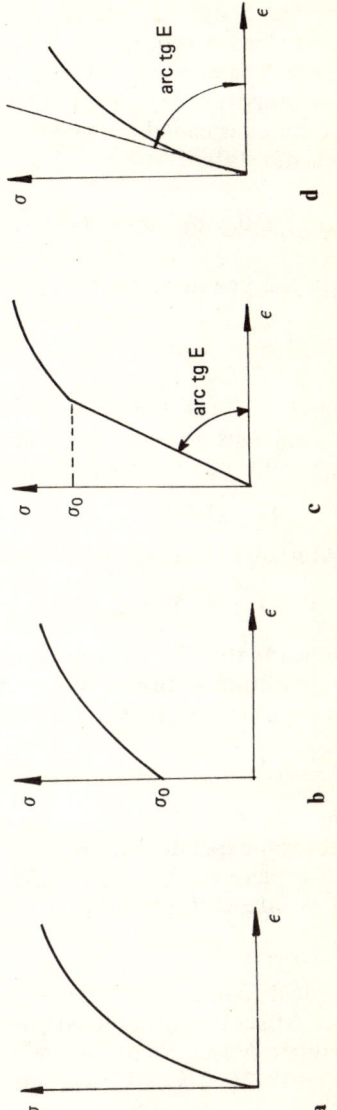

Figure 7.2a–d. Nonlinear idealizations of work-hardening materials. **a** Work-hardening plastic material. **b** Work-hardening rigid-plastic material. **c** Work-hardening elasto-plastic material. **d** Elasto-plastic Ramberg–Osgood material.

7.2 Ideal Plasticity Theories

7.2.1 Yield Criteria

In order to formulate a transition criterion of isotropic ideal plastic material from elastic state to plastic state, one assumes the existence of an invariant scalar-valued function $f(\sigma_{ij})$ which can be used to formulate the effort criterion and which is constant on the boundary of the domain corresponding to the elastic range, i.e. $f(\sigma_{ij}) = c$. The above condition defines a hyperplane in stress space which is called the yield surface. For any function $F = f - c$ material is in the elastic state if

$$F(\sigma_{ij}) < 0, \quad \text{or} \quad F(\sigma_{ij}) = 0 \quad \text{and} \quad \frac{\partial F}{\partial \sigma_{ij}} \dot{\sigma}_{ij} < 0 \qquad (7.3)$$

and in the plastic state if

$$F(\sigma_{ij}) = 0 \quad \text{and} \quad \frac{\partial F}{\partial \sigma_{ij}} \dot{\sigma}_{ij} = 0. \qquad (7.4)$$

States $F(\sigma_{ij}) > 0$ in ideal plasticity theory are not admissible. If we limit the considerations to isotropic materials, then in place of the general function $F(\sigma_{ij}) = 0$ we can put a symmetric function of principal stresses

$$F(\sigma_1, \sigma_2, \sigma_3) = 0, \qquad (7.5)$$

or a function of stress invariants, for instance

$$F(J_{1\sigma}, J_{2\sigma}, J_{3\sigma}) = 0. \qquad (7.6)$$

This means that for isotropic material the yield condition does not depend on the orientation of principal directions. For determination of the function F we consider in many cases another system of invariants

$$F(J_{1\sigma}, J_{2s}, J_{3s}) = 0. \qquad (7.7)$$

The majority of experiments carried out for metals indicate that the yield surface is independent of the mean hydrostatic pressure, i.e. of the $J_{1\sigma}$ invariant. Then in place of the general expression (7.7) we get the equation of a tubular yield surface with the tube axis equally inclined to $\sigma_1, \sigma_2, \sigma_3$

$$F(J_{2s}, J_{3s}) = 0. \qquad (7.8)$$

The yield conditions, based on the tubular yield surface equation are: Huber–Mises condition (HM), and Treska–Guest condition (TG). According to the Huber–Mises condition the criterion of material transition in plastic state depends on the second invariant only, that is

$$F(J_{2s}) = \bar{\sigma}_1 - \sigma_0 = \sqrt{-3J_{2s}} - \sigma_0 = \sqrt{\tfrac{3}{2} s_{ij} s_{ij}} - \sigma_0 = 0, \qquad (7.9)$$

or, in component form

$$(\sigma_x - \sigma_y)^2 + (\sigma_y - \sigma_z)^2 + (\sigma_z - \sigma_x)^2 + 6(\tau_{xy}^2 + \tau_{yz}^2 + \tau_{zx}^2) - 2\sigma_0^2 = 0. \qquad (7.10)$$

In a physical sense the HM hypothesis means that the energy of non-dilatational deformation $\phi = \frac{1}{2}s_{ij}e_{ij} = s_{ij}s_{ij}/4G$ depends upon plasticization.

In the space of principal stresses $\sigma_1, \sigma_2, \sigma_3$, the yield surface HM represents a cylinder with axis $\sigma_1 = \sigma_2 = \sigma_3$. According to Treska–Guest, the transition criterion is assumed to be the magnitude of the vector of the largest tangential stress $\tau_{max} = \frac{1}{2}(\sigma_I - \sigma_{III})$. In symmetrical and invariant form, the TG condition has the form

$$[(\sigma_1 - \sigma_2)^2 - \sigma_0^2][(\sigma_2 - \sigma_3)^2 - \sigma_0^2][(\sigma_3 - \sigma_1)^2 - \sigma_0^2] = 0, \quad (7.11)$$

or, in terms of the invariants J_{2s}, J_{3s}

$$4J_{2s}^3 - 27J_{3s}^2 - 9\sigma_0^2 J_{2s}^2 + 6\sigma_0^4 J_{2s} - \sigma_0^6 = 0. \quad (7.12)$$

In the space of principal stresses $\sigma_1, \sigma_2, \sigma_3$, the yield surface TG represents a prism of hexagonal cross-section, inscribed in the cylinder HM. Cross-sections of both the surfaces are presented in Fig. 7.3. There are several other yield criteria available in the literature. Among generalizations of the ideal plasticity conditions we mention only the criterion due to von Mises (1928) which can be formulated as follows:

$$\Pi_{ijkl}\sigma_{ij}\sigma_{kl} = 1, \quad \Pi_{ijkl} = \Pi_{klij} = \Pi_{jikl} = \Pi_{ijlk} \quad (7.13)$$

where Π_{ijkl} are the plasticity tensor (fourth order) moduli. Eq. (7.13) takes the form

$$\begin{aligned}
&\Pi_{xxxx}\sigma_x^2 + \Pi_{yyyy}\sigma_y^2 + \Pi_{zzzz}\sigma_z^2 + 2\Pi_{xxyy}\sigma_x\sigma_y + 2\Pi_{yyzz}\sigma_y\sigma_z \\
&+ 2\Pi_{zzxx}\sigma_z\sigma_x + 4\Pi_{xxxy}\sigma_x\tau_{xy} + 4\Pi_{xxyz}\sigma_x\tau_{yz} + 4\Pi_{xxzx}\sigma_x\tau_{zx} \\
&+ 4\Pi_{yyxy}\sigma_y\tau_{xy} + 4\Pi_{yyyz}\sigma_y\tau_{yz} + 4\Pi_{yyzx}\sigma_y\tau_{zx} + 4\Pi_{zzxy}\sigma_z\tau_{xy} \\
&+ 4\Pi_{zzyz}\sigma_z\tau_{yz} + 4\Pi_{zzzx}\sigma_z\tau_{zx} + 8\Pi_{xyyz}\tau_{xy}\tau_{yz} + 8\Pi_{yzzx}\tau_{yz}\tau_{zx} \\
&+ 8\Pi_{zxxy}\tau_{zx}\tau_{xy} + 4\Pi_{xyxy}\tau_{xy}^2 + 4\Pi_{yzyz}\tau_{yz}^2 + 4\Pi_{zxzx}\tau_{zx}^2 = 1.
\end{aligned} \quad (7.14)$$

7.2.2 Hencky–Iljuszyn (HI) Deformation Theory

Deformation theory, also termed the theory of small elasto-plastic deformations, was first used by Nadai (1923), then formulated by Hencky (1924) and developed by Iljuszyn (1943, 1948). In contrast with the other plasticity theories it is based on the assumption that there exists a certain relationship between stress tensor and strain tensor (not strain rate tensor) which may be regarded as a generalization of physical relationships which characterize the nonlinear elasticity theory. As the basis of the theory one states the following postulates:

1. The principal directions of the stress tensor are identical with the principal directions of the strain tensor
2. Mean stress σ_m is proportional to the mean strain, and the coefficient of proportionality is the same as in the law of volume change in elasticity theory
3. Effective stress $\bar{\sigma}$ is a certain function of effective strain $\bar{\varepsilon}$, which is to be determined experimentally

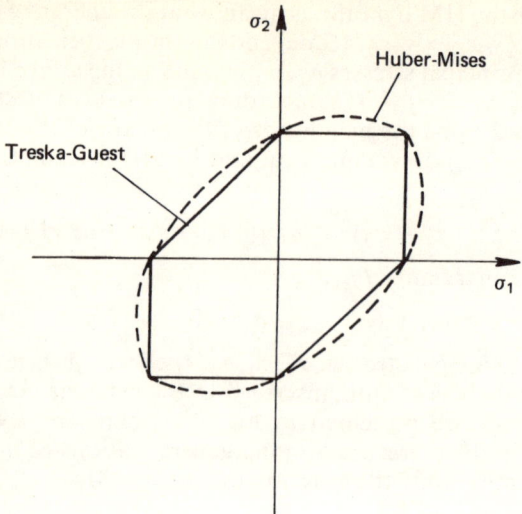

Figure 7.3. Ideal plasticity conditions.

Then the constitutive equations for active plastic processes have the forms

$$e_{ij} = \varphi s_{ij}, \qquad \varepsilon_{kk} = \frac{1-2\nu}{E} \sigma_{kk}, \tag{7.15}$$

where the function $\varphi = \varphi(\bar{\varepsilon})$ ($\varphi > 1/2G$) determines promotion of plastic strains; in ideal plasticity theory this function can be determined from the yield condition. Multiplying the first of the Eqs. (7.15) by itself and using definitions of $\bar{\varepsilon}$ and $\bar{\sigma}$ we get for the HM condition ($\bar{\sigma} = \sigma_0$)

$$\varphi = \tfrac{3}{2}\frac{\bar{\varepsilon}}{\bar{\sigma}(\bar{\varepsilon})} = \tfrac{3}{2}\frac{\bar{\varepsilon}}{\sigma_0} = \frac{\sqrt{\tfrac{3}{2} e_{ij} e_{ij}}}{\sigma_0}. \tag{7.16}$$

For description of passive processes we take an equation of different form to that for nonlinear elasticity,

$$\tilde{e}_{ij} - e_{ij} = \frac{1}{2G}(\tilde{s}_{ij} - s_{ij}) \tag{7.17}$$

where $\sim$ refers to the point at which the passive process starts. The equation of the boundary between active and passive processes is obtained by requiring that the equation of passive processes and the HM condition are satisfied simultaneously. We get in such a way the equation of neutral processes starting at point $(\tilde{s}_{ij}, \tilde{e}_{ij})$

$$G(\tilde{e}_{ij} - e_{ij})(\tilde{e}_{ij} - e_{ij}) - \tilde{s}_{ij}(\tilde{e}_{ij} - e_{ij}) = 0. \tag{7.18}$$

In terms of principal components this equation can be rewritten as

$$G[(\tilde{e}_1 - e_1)^2 + (\tilde{e}_2 - e_2)^2 + (\tilde{e}_3 - e_3)^2] \\ - \tilde{s}_1(\tilde{e}_1 - e_1) - \tilde{s}_2(\tilde{e}_2 - e_2) - \tilde{s}_3(\tilde{e}_3 - e_3) = 0,$$ (7.19)

where $e_j = \varepsilon_j - \varepsilon_m$, $s_j = \sigma_j - \sigma_m$ $(j = 1, 2, 3)$.

7.2.3 Plastic Flow Theory

In plastic flow theories, called also incremental theories, we formulate physical equations in the form of relationships between the strain rate tensor $\dot{\varepsilon}_{ij} = d\varepsilon_{ij}/dt$ and the stress tensor. Moreover one assumes that the change of material volume takes place according to the laws of the theory of elasticity. Then Eq. (7.15) no longer applies and the law governing the change of shape can be expressed as

$$\dot{e}_{ij}^p = \lambda s_{ij}$$ (7.20)

where, on the basis of assumption of small strains, $\dot{e}_{ij}^p = \dot{e}_{ij} - \dot{e}_{ij}^e$.

Assume an ideal rigid-plastic material model. Then we get, for active processes, the equation of Levy–Mises theory

$$\dot{e}_{ij} = \lambda s_{ij}.$$ (7.21)

A passive process refers to the rigid behaviour of material

$$\dot{e}_{ij} = 0.$$ (7.22)

The modulus λ can be eliminated on the basis of yield condition (HM for instance). Using definitions

$$\dot{\bar{\varepsilon}} = \sqrt{\tfrac{2}{3} \dot{e}_{ij} \dot{e}_{ij}}, \qquad \bar{\sigma} = \sqrt{\tfrac{2}{3} s_{ij} s_{ij}}$$ (7.23)

we finally get

$$\lambda = \frac{3}{2} \frac{\dot{\bar{\varepsilon}}}{\sigma_0},$$ (7.24)

or, taking into account that for rigid-plastic material $\dot{\bar{\varepsilon}} = \dot{\bar{\varepsilon}}^p$,

$$\lambda^{LM} = \frac{3}{2} \frac{s_{ij} \dot{e}_{ij}}{\sigma_0^2} = \frac{3}{2} \frac{\dot{W}^p}{\sigma_0^2}.$$ (7.25)

Assuming the model of ideal elasto-plastic material and using for elastic deformations Hooke's law

$$e_{ij}^e = \frac{1}{2G} s_{ij}$$ (7.26)

we get, for active processes, the equation of Prandtl–Reuss theory

$$\dot{e}_{ij} = \frac{\dot{s}_{ij}}{2G} + \lambda s_{ij}, \qquad F(\sigma_{ij}) = 0,$$ (7.27)

and for passive processes

$$\dot{e}_{ij} = \frac{\dot{s}_{ij}}{2G}, \qquad F(\sigma_{ij}) < 0. \tag{7.28}$$

The modulus λ one obtains by multiplication of both sides of Eq. (7.21) by s_{ij}

$$s_{ij}\dot{e}_{ij} = \frac{1}{2G} s_{ij}\dot{s}_{ij} + \lambda s_{ij} s_{ij}, \tag{7.29}$$

and making use of the fact that for ideal plasticity described by HM condition $s_{ij}s_{ij} = \frac{2}{3}\sigma_0^2$, the first term in the right-hand side of Eq. (7.29) vanishes, $s_{ij}\dot{s}_{ij} = 0$. Then λ is finally determined by the speed of plastic strain dissipation

$$\lambda^{PR} = \frac{3}{2} \frac{s_{ij}\dot{e}_{ij}}{\sigma_0^2} = \frac{3}{2} \frac{\dot{W}^p}{\sigma_0^2}, \tag{7.30}$$

similar to Levy–Mises theory, Eq. (7.25). However Eq. (7.24) does not hold. Multiplying Eq. (7.27) by dt leads to the Prandtl–Reuss equation in incremental form

$$de_{ij} = \frac{1}{2G} ds_{ij} + s_{ij} d\lambda. \tag{7.31}$$

In engineering notation Eq. (7.31) can be expressed in the following way

$$de_x = \frac{ds_x}{2G} + s_x d\lambda, \qquad d\gamma_{xy} = \frac{d\tau_{xy}}{G} + 2\tau_{xy} d\lambda,$$

$$de_y = \frac{ds_y}{2G} + s_y d\lambda, \qquad d\gamma_{yz} = \frac{d\tau_{yz}}{G} + 2\tau_{yz} d\lambda, \tag{7.32}$$

$$de_z = \frac{ds_z}{2G} + s_z d\lambda, \qquad d\gamma_{zx} = \frac{d\tau_{zx}}{G} + 2\tau_{zx} d\lambda,$$

where

$$d\lambda = \frac{3}{2\sigma_0^2} (s_x de_x + s_y de_y + s_z de_z + \tau_{xy} d\gamma_{xy} + \tau_{yz} d\gamma_{yz} + \tau_{zx} d\gamma_{zx}) \tag{7.33}$$

7.2.4 Comparison of Flow Theory and Deformation Theory

Flow theory and deformation theory can be obtained as particular cases of the general linear law of Hohenemser and Prager (1932)

$$\alpha_1 s_{ij} + \alpha_2 \dot{s}_{ij} + \alpha_3 e_{ij} + \alpha_4 \dot{e}_{ij} = 0, \tag{7.34}$$

in which three of the coefficients α_j ($j = 1, 2, 3, 4$) are constant and one is a function of strain. The more general theory presented by Reiner (1945) is a generalization of the plastic flow theory and is based on the equation

$$\sigma_{ij} = \alpha_0 \delta_{ij} + \alpha_1 \dot{\varepsilon}_{ij} + \alpha_2 \dot{\varepsilon}_{ik} \dot{\varepsilon}_{kj}, \tag{7.35}$$

where $\dot{\varepsilon}_{ij}$ is the tensor of total or plastic strain rate, for rigid-plastic or elasto-plastic material, respectively.

In the flow theory a current state is dependent on loading history. This is no longer the case in the deformation theory. In order to show conditions when the above theories give the same results we differentiate Eq. (7.15)

$$\dot{e}_{ij} = \dot{\varphi} s_{ij} + \varphi \dot{s}_{ij}. \tag{7.36}$$

Multiplying the above equation and the Prandtl–Reuss equation by s_{ij}, making use of the HM condition, $s_{ij}\dot{s}_{ij} = 0$, and next comparing the left-hand sides of the two equations, we find the condition $\dot{\varphi} - \lambda = 0$. Eq. (7.36) can be now expressed in the form

$$\dot{e}_{ij} = \lambda s_{ij} + \varphi \dot{s}_{ij}. \tag{7.37}$$

These equations give the same results as Prandtl–Reuss theory only in the case when in an active process the point represented by stress state in stress space is fixed

$$s_{ij} = s_{ij}^0 \qquad (\dot{s}_{ij} = 0). \tag{7.38}$$

After integrating the plastic flow equation (7.27) we get the condition of proportional increase of deviatoric strain components

$$e_{ij} = e_{ij}^0 f(t). \tag{7.39}$$

A process for which deviatoric strain components increase proportionally is called a simple process. Iljuszyn has formulated the theorems governing simple loading. If external loading increases proportionally, then deviatoric strain components in each point of the body also increase proportionally and the process is simple; then results of three classical plasticity theories are convergent. The Iljuszyn theorem is fulfilled with the following major assumptions: the material is incompressible, isotropic, homogeneous and work-hardening (geometric effects are omitted). In the case of non-simple loading the Hencky–Iljuszyn theory gives false results. The Hencky–Iljuszyn theory leads to internal inconsistency for description of neutral processes.

For neutral processes we need to satisfy the equations of active and passive processes (Figs. 7.4 and 7.5). By deformation theory

$$e_{ij} = \varphi s_{ij}, \qquad \tilde{e}_{ij} - e_{ij} = \frac{1}{2G}(\tilde{s}_{ij} - s_{ij}), \qquad s_{ij}s_{ij} = \tfrac{2}{3}\sigma_0^2. \tag{7.40}$$

Differentiating two of the above conditions and making use of the third one, we are led to the following continuity condition for the description of neutral processes

$$\left(\varphi - \frac{1}{2G}\right)\dot{s}_{ij} = 0. \tag{7.41}$$

Unless $\varphi = \tfrac{1}{2}G$ this condition represents a contradiction. On the basis of Prandtl–Reuss equations (describing neutral processes) we get

$$\dot{e}_{ij} = \frac{1}{2G}\dot{s}_{ij} + \lambda s_{ij}, \qquad \dot{e}_{ij} = \frac{1}{2G}\dot{s}_{ij}, \qquad s_{ij}\dot{s}_{ij} = 0. \tag{7.42}$$

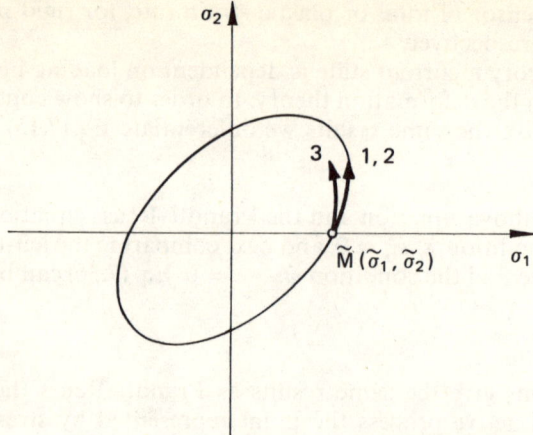

Figure 7.4. Active (1), neutral (2) and passive (3) processes in stress space.

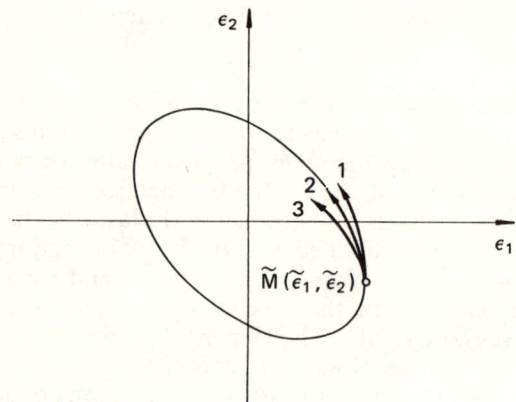

Figure 7.5. Active (1), neutral (2) and passive (3) processes in strain space.

The continuity condition for neutral processes

$$\lambda = \frac{3}{2} \frac{s_{ij} \dot{e}_{ij}}{\sigma_0^2} = 0 \tag{7.43}$$

is fulfilled, because for these processes no dissipation of energy occurs.

7.2.5 Ideal Plasticity Theory for Finite Deformations

The equations of ideal plasticity theory for finite deformations have simple form only in the case of nonrotational deformations. In this case we can formulate convenient generalizations of classical equations of ideal plasticity theory for principal logarithmic strains and principal real stresses. These theories with respect to rigid-plastic materials with hardening have been formulated by Nadai (1937a,b, 1950) and Davis (1937, 1943). For ideal plastic materials the generality of the HM condition on Cauchy stresses is valid

$$s_J s_J = \tfrac{2}{3}\sigma_0^2. \tag{7.44}$$

The Nadai–Davis deformation theory assumes the similarity of principal deviatoric logarithmic strains e_J^L and principal real stresses (Cauchy) s_J in a fixed material particle X_I. Then

$$e_J^L = \phi s_J, \quad \phi = \frac{3}{2}\frac{\bar{\varepsilon}^L}{\sigma_0} = \frac{\sqrt{\tfrac{3}{2} e_J^L e_J^L}}{\sigma_0} \quad (J = 1,2,3). \tag{7.45}$$

The passive processes are described by respective generalizations of Eq. (7.17)

$$\tilde{e}_J^L - e_J^L = \frac{1}{2G}(\tilde{s}_J - s_J) \quad (J = 1,2,3). \tag{7.46}$$

Generalization of the Levy–Mises equation theory leads to an incremental Nadai–Davis theory

$$de_J^L = \Lambda s_J, \quad \Lambda = \frac{3}{2}\frac{d\bar{\varepsilon}^L}{\sigma_0} = \frac{\sqrt{\tfrac{3}{2} de_J^L de_J^L}}{\sigma_0} \quad (J = 1,2,3). \tag{7.47}$$

The symbol $d\bar{\varepsilon}^L$ represents the effective increments of principal logarithmic strains in a fixed material particle. Then by using the material derivative we get

$$\dot{e}_J^L = \Lambda s_J, \quad \Lambda = \frac{3}{2}\frac{\dot{\bar{\varepsilon}}^L}{\sigma_0} = \frac{\sqrt{\tfrac{3}{2} \dot{e}_J^L \dot{e}_J^L}}{\sigma_0} \quad (J = 1,2,3). \tag{7.48}$$

Generalization of Prandtl–Reuss equation theory needs the hypothesis of decomposition of total rates of logarithmic strains into elastic and plastic parts. Assuming the additive decomposition

$$\dot{e}_J^L = \dot{e}_J^{Le} + \dot{e}_J^{Lp} \tag{7.49}$$

and applying the logarithmic Hooke's law to the first part

$$\dot{e}_J^{Le} = \frac{1}{2G}\dot{s}_J \quad (J = 1,2,3) \tag{7.50}$$

and the incremental Nadai–Davis relation to the plastic part

$$\dot{e}_J^{Lp} = \Lambda s_J \quad (J = 1,2,3) \tag{7.51}$$

we finally obtain

$$\dot{e}_J^L = \frac{1}{2G}\dot{s}_J + \psi \Lambda s_J \quad (J = 1,2,3) \tag{7.52}$$

$$\Lambda = \frac{3}{2}\frac{\dot{\bar{\varepsilon}}^{Lp}}{\sigma_0} = \frac{\sqrt{\tfrac{3}{2}\dot{e}_J^{Lp}\dot{e}_J^{Lp}}}{\sigma_0} \tag{7.53}$$

where the symbol $\dot{\bar{\varepsilon}}^{Lp}$ represents the effective logarithmic plastic strain rate, $\dot{\bar{\varepsilon}}^{Lp} = \sqrt{(\tfrac{3}{2}\dot{e}_I^{Lp}\dot{e}_I^{Lp})}$, and the coefficient ψ takes the value 1 for active processes and 0 for passive processes.

8 Work-Hardening Equation

8.1 Drucker Postulate

8.1.1 Stability of Plastic Material in the Drucker Sense

Following Drucker (1951, 1959), we introduce the concept of stability of deformation processes. We use the so-called quasi-cycle for stresses, e.g. a process in which the stress after loading and unloading returns to its initial state, but the strain state in general does not (through irreversibility of plastic processes). Fig. 8.1 shows the quasi-cycle for elasto-plastic material in the systems (σ, ε) and (σ, ε^p) (line $(M_0 M_1 M_2 M_3)$. If the point representing the initial stress state lies in the elastic range, and during the quasi-cycle reaches a yield surface, then elementary work done by external factors on this cycle corresponding to infinitesimal plastic strain is

$$dW^p = (\sigma - \sigma^0)d\varepsilon^p + \tfrac{1}{2}d\sigma\, d\varepsilon^p + \ldots \tag{8.1}$$

During a deformation process a material is said to be stable in the Drucker sense, if for each arbitrary initial state σ_{ij}^0, lying inside the yield surface, the work done during a quasi-cycle is positive (Fig. 8.2)

$$dW^p > 0. \tag{8.2}$$

The relation (8.2) is fulfilled if

$$(\sigma_{ij} - \sigma_{ij}^0)d\varepsilon_{ij}^p \geq 0, \tag{8.3}$$

for an initial point lying inside the yield surface ($\sigma_{ij} \neq \sigma_{ij}^0$), or

$$d\sigma_{ij}d\varepsilon_{ij}^p > 0, \tag{8.4}$$

for an initial point lying on the yield surface ($\sigma_{ij} = \sigma_{ij}^0$).

If a deformation process satisfies a weaker condition

$$dW^p \geq 0, \tag{8.5}$$

then the process is said to be neutrally stable in the Drucker sense. Ideal plastic material is neutral in the Drucker sense, because

$$(\sigma_{ij} - \sigma_{ij}^0)d\varepsilon_{ij}^p \geq 0, \tag{8.5a}$$

$$d\sigma_{ij}d\varepsilon_{ij}^p = 0. \tag{8.5b}$$

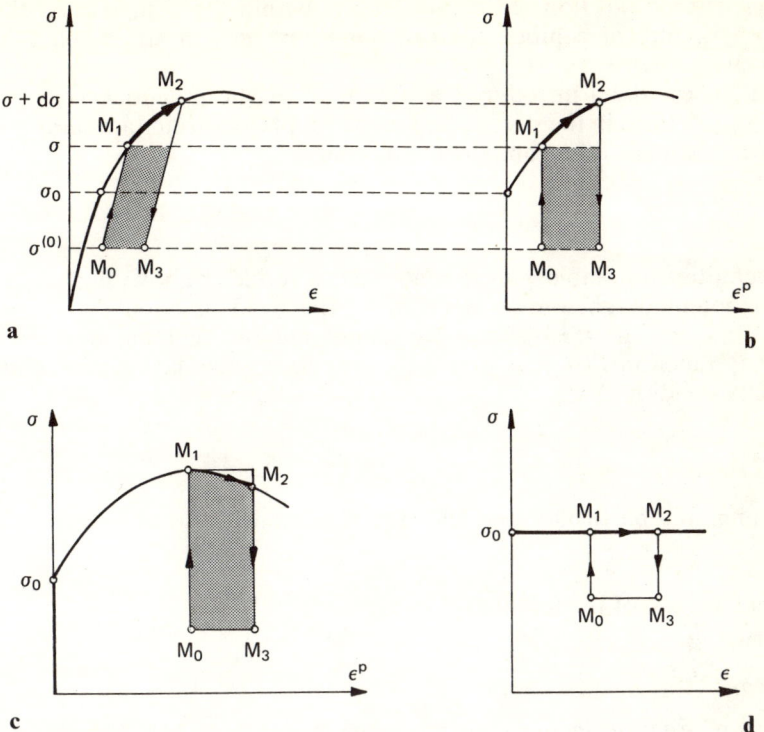

Figure 8.1a–d. Drucker postulate. **a** Stable material. **b** Unstable material. **c** Neutral material. **d** Neutrally stable material.

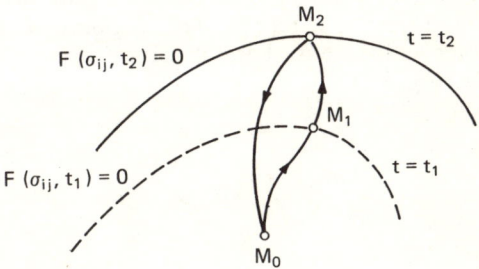

Figure 8.2. Quasi-cycle in stress space.

8.1.2 Associated Plastic Flow

The Drucker postulate implies the following conclusions: (1) for stable material the yield surface is convex, (2) the vector of increase of the plastic strain, $d\varepsilon_{ij}^p$, is normal to the yield surface and directed outwards from the surface. In reality, for each position of the initial point within the yield surface, the angle between vectors of components $(\sigma_{ij} - \sigma_{ij}^0)$ and $d\varepsilon_{ij}^p$ in stress space is acute (Fig. 8.3).

Von Mises (1928) introduced a notion of plastic potential, i.e. a function $g = g(\sigma_{ij})$, defined in the stress space such that plastic strain rate components $\dot{\varepsilon}_{ij}^p$ are proportional to the components of grad g

$$\dot{\varepsilon}_{ij}^p = \lambda \frac{\partial g(\sigma_{ij})}{\partial \sigma_{ij}}. \tag{8.6}$$

The definition of stable plastic materials introduced above implies that the plastic potential function is identical to the yield function determining the convex surface in stress space. By identifying the function $g(\sigma_{ij})$ with the plasticity function $F(\sigma_{ij})$ we derive the associated flow law (with the assumed plasticity condition) as

$$\dot{\varepsilon}_{ij}^p = \lambda \frac{\partial F(\sigma_{ij})}{\partial \sigma_{ij}}. \tag{8.7}$$

Assuming, in a particular case, the HM condition

$$F(\sigma_{ij}) = s_{ij} s_{ij} - \tfrac{2}{3} \sigma_0^2 \tag{8.8}$$

and making use of the fact that

$$\frac{\partial F^{HM}(\sigma_{ij})}{\partial \sigma_{ij}} = s_{ij}, \tag{8.9}$$

we get an equation identical to the Prandtl–Reuss equation for ideal plastic material

$$\dot{\varepsilon}_{ij}^p = \lambda s_{ij}, \tag{8.10}$$

but referred to the current configuration of an arbitrary elasto-plastic medium.

If the stress state at time t_1 is represented by point M_1 lying on the current yield surface, then for active, neutral and passive processes starting at this point we get respectively (Fig. 8.4)

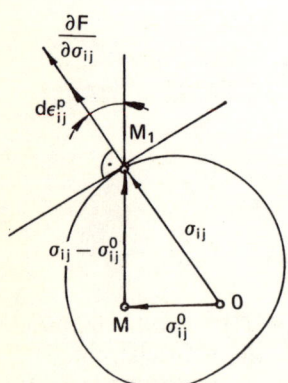

Figure 8.3. Conclusions from the Drucker postulate.

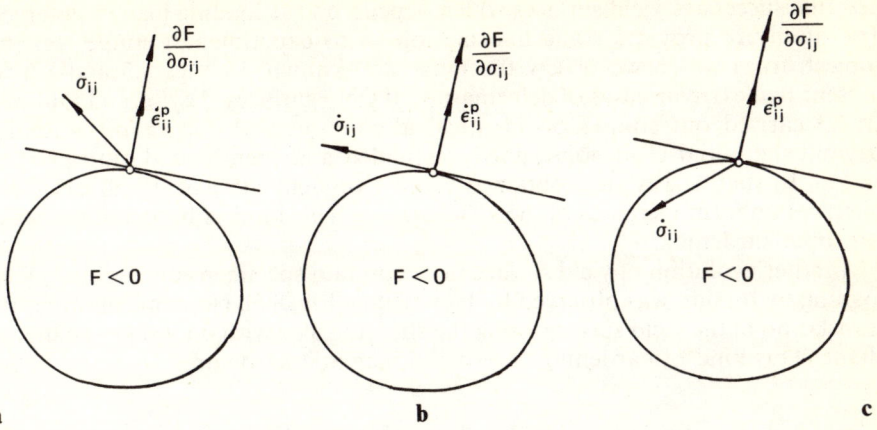

Figure 8.4. Vectors of stress increments for **a** active, **b** neutral and **c** passive processes.

$$\frac{\partial F}{\partial \sigma_{ij}} \dot{\sigma}_{ij} \begin{cases} > 0 & \text{active,} \\ = 0 & \text{neutral,} \\ < 0 & \text{passive.} \end{cases} \qquad (8.11)$$

In the case of an ideal plastic material an exit beyond the yield surface is impossible, then in order to separate neutral and active processes we should examine a dissipation rate $\dot{W}^p$, which on the basis of associated plastic flow is given by

$$\dot{W}^p = \sigma_{ij}\dot{\varepsilon}^p_{ij} = \lambda\sigma_{ij}\frac{\partial F(\sigma_{ij})}{\partial \sigma_{ij}}. \qquad (8.12)$$

From the assumption of convexity of the yield surface it follows that $\partial F/\partial \sigma_{ij}(\sigma_{ij}) > 0$. Then for active processes $\lambda > 0$, and for neutral and passive processes $\lambda = 0$. Finally for an ideal plastic material and a point lying on the yield surface the following may hold

$$\lambda \begin{cases} > 0 \\ = 0 \\ = 0 \end{cases} \quad \text{and} \quad \frac{\partial F}{\partial \sigma_{ij}} \dot{\sigma}_{ij} \begin{cases} > 0 & \text{active,} \\ = 0 & \text{neutral,} \\ < 0 & \text{passive.} \end{cases} \qquad (8.13)$$

8.2 Yield Surfaces for Work-Hardening Materials

8.2.1 Experimental Results

The fundamental assumption in the ideal plasticity theory is that the yield surface does not change during the loading process. In practice in many materials one observes deformation of the yield surface after unloading. To take account of this effect equations for the initial yield surface have to be modified to describe successive yield surfaces which depend on the loading history assumed. The literature provides some information as to experimental studies on this subject, from which we review the most well-known. In Figs. 8.5 and 8.6 we present two extreme cases of deformation of yield surfaces. Taylor and Quinney (1931) carried out studies on identical aluminium and copper pipes loaded beyond the initial yield point, partly unloaded and then loaded until plasticization. In such a way they obtained successive yield surfaces. In all cases one observes uniform increases of the yield surface. This kind of hardening is called isotropic hardening.

Another evolution of yield surface in aluminium specimens undergoing cyclic loading by torsion was observed by Ivey (1961) (Fig. 8.6). Here one observes the translation of the yield surface along the shearing axis with only slight change of shape. This kind of hardening is termed kinematic hardening.

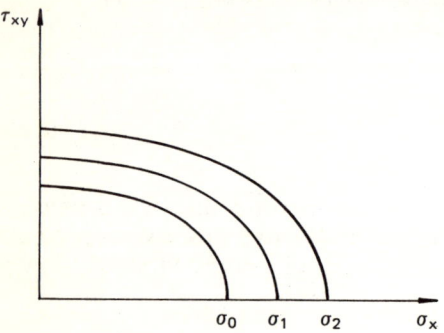

Figure 8.5. Evolution of yield surface (after Taylor and Quinney).

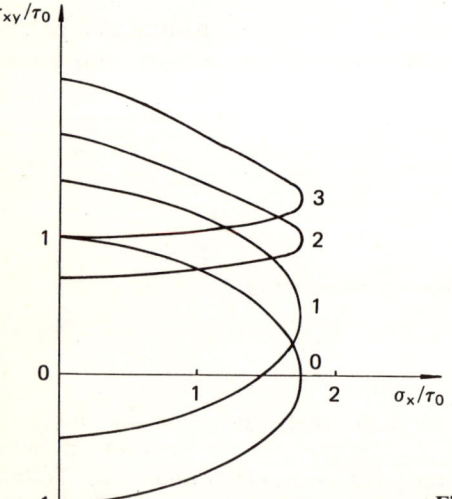

Figure 8.6. Evolution of yield surface (after Ivey).

8.2.2 Isotropic Hardening

The basic hypothesis of isotropic hardening is the assumption that the shape of the yield surface is unchanging, and its growth can be described by one scalar parameter which is a function of plastic deformation. Figure 8.7 illustrates the principle of isotropic hardening of material. The material is first uniaxially loaded beyond its initial yield strength σ_0^0, to an instantaneous strain ε_1, followed by unloading upon reaching point M. A permanent strain ε_2 is expected to be introduced in the material after the completion of the unloading. If the material is loaded again, it is found to yield at a higher yield strength σ_0^1 which coincides with the stress at the last loading level.

We can describe the plastic potential function F by the equation

$$F = F(\sigma_{ij}, K, T, \dot{\bar{\varepsilon}}) \tag{8.14}$$

where K is the work-hardening parameter.

The Odqvist hypothesis (1933) assumes

$$dK = \sqrt{d\varepsilon_{ij}^p d\varepsilon_{ij}^p} \tag{8.15}$$

According to the Quinney–Taylor hypothesis

$$dK = \sigma_{ij} d\varepsilon_{ij}^p. \tag{8.16}$$

This kind of hardening is called energetic hardening. If we differentiate F by using the chain rule for partial differentiation we obtain

$$\dot{F} = \frac{\partial F}{\partial \sigma_{ij}} \dot{\sigma}_{ij} + \frac{\partial F}{\partial K} \dot{K} + \frac{\partial F}{\partial T} \dot{T} + \frac{\partial F}{\partial \dot{\bar{\varepsilon}}} (\dot{\bar{\varepsilon}})^{\cdot}. \tag{8.17}$$

Using the Quinney–Taylor hypothesis, it is seen that the second term in the right-hand side of (8.17) can be expressed in terms of plastic strains $\dot{\varepsilon}_{ij}^p$ as

$$\frac{\partial F}{\partial K} \dot{K} = \frac{\partial F}{\partial K} \frac{\partial K}{\partial \varepsilon_{ij}^p} \dot{\varepsilon}_{ij}^p. \tag{8.18}$$

Combining (8.17) and (8.18) gives

$$\dot{F} = \frac{\partial F}{\partial \sigma_{ij}} \dot{\sigma}_{ij} + \frac{\partial F}{\partial K} \frac{\partial K}{\partial \varepsilon_{ij}^p} \dot{\varepsilon}_{ij}^p + \frac{\partial F}{\partial T} \dot{T} + \frac{\partial F}{\partial \dot{\bar{\varepsilon}}} (\dot{\bar{\varepsilon}})^{\cdot}. \tag{8.19}$$

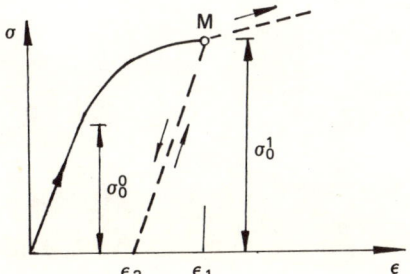

Figure 8.7. Loading and unloading processes for isotropic hardening.

Equilibrium conditions for small increments of plastic deformation require that variation of the plastic energy be stationary:

$$\dot{F} = \frac{\partial F}{\partial \sigma_{ij}} \dot{\sigma}_{ij} + \frac{\partial F}{\partial K} \frac{\partial K}{\partial \varepsilon_{ij}^p} \dot{\varepsilon}_{ij}^p + \frac{\partial F}{\partial T} \dot{T} + \frac{\partial F}{\partial \bar{\varepsilon}} (\dot{\bar{\varepsilon}})^{\cdot} = 0. \tag{8.20}$$

8.2.3 Kinematic Hardening

In 1881 Bauschinger observed in plastic deformation processes a phenomenon of increase of the yield point in tension and decrease in compression (with respect to initial value σ_0). This effect is illustrated in Fig. 8.8. This behaviour is called a kinematic hardening or Bauschinger effect. In this kind of hardening one observes the unchanging shape of the yield surface with its translation. An illustration of kinematic hardening for a state of biaxial stress is shown in Fig. 8.9. As can be seen from this figure, the loading curve shifts by an amount α_{ij} upon a change of stress state from 1 to 2. For cases involving significant temperature variation, the yield function and flow curves become temperature

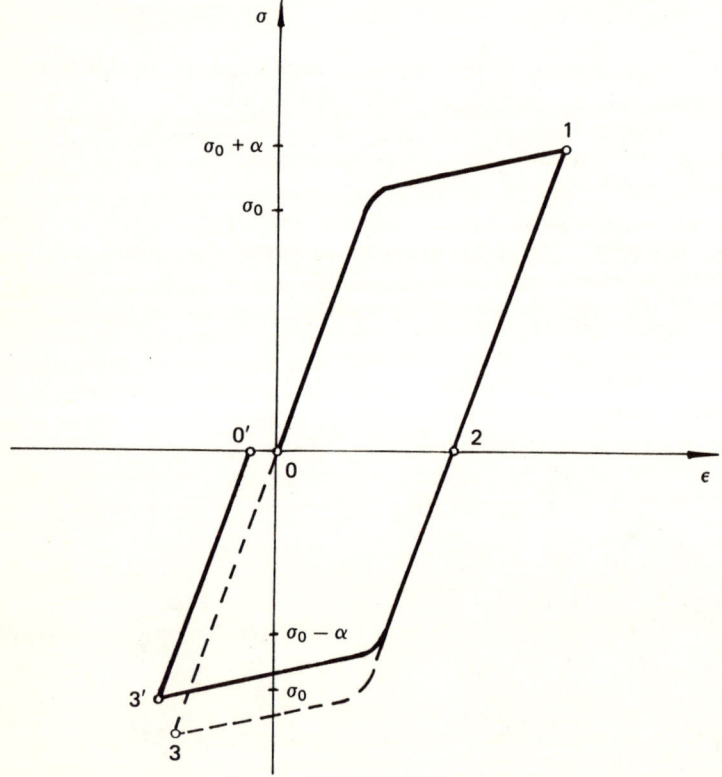

Figure 8.8. Bauschinger effect.

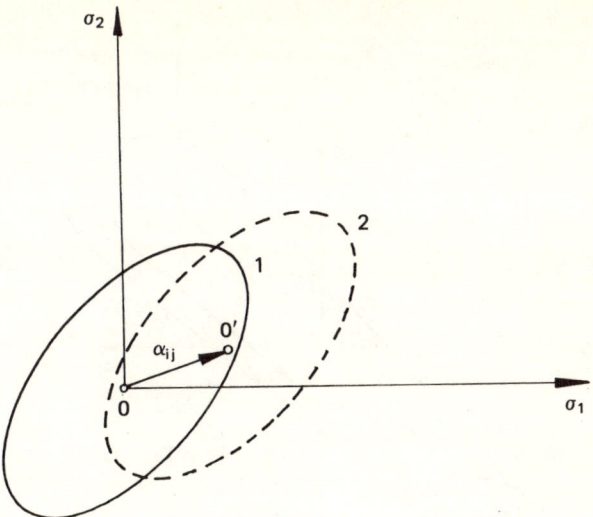

Figure 8.9. Loading path for kinematic hardening.

dependent. The change of the yield surface due to a temperature increment is illustrated in Fig. 8.10.

For a structure subject to combined cyclic thermal and mechanical loadings, both types of loading path variations have to be included. No rigid rule on the proper sequence of the shift and expansion (or contraction) of the yield surfaces is necessary, as the effects of both these variations in the mathematical model are additive as will be shown in the next derivations. The original kinematic hardening theory proposed by Prager was based on the assumption that translation increments of the loading surface in the nine-dimensional stress space take place in the direction of the outwardly directed normal to the surface at an instantaneous stress state. In processes of kinematic hardening, in order to describe the motion of the initial yield surface, one introduces the translation tensor α_{ij}, whose components determine a new position of the yield surface centre.

Ziegler (1959) assumes a motion of the surface in the direction of the difference of σ and α.

$$d\alpha_{ij} = d\mu(\sigma_{ij} - \alpha_{ij}) \tag{8.21}$$

where $d\mu$ is the multiplier.

Melan (1938) has proposed definition of the tensor α_{ij} as

$$d\alpha_{ij} = C d\varepsilon_{ij}^p. \tag{8.22}$$

This definition was also used by Prager (1955).

Considering the modification of the yield surface with respect to temperature and strain rate changes we have

$$C = C(T, \dot{\varepsilon}). \tag{8.23}$$

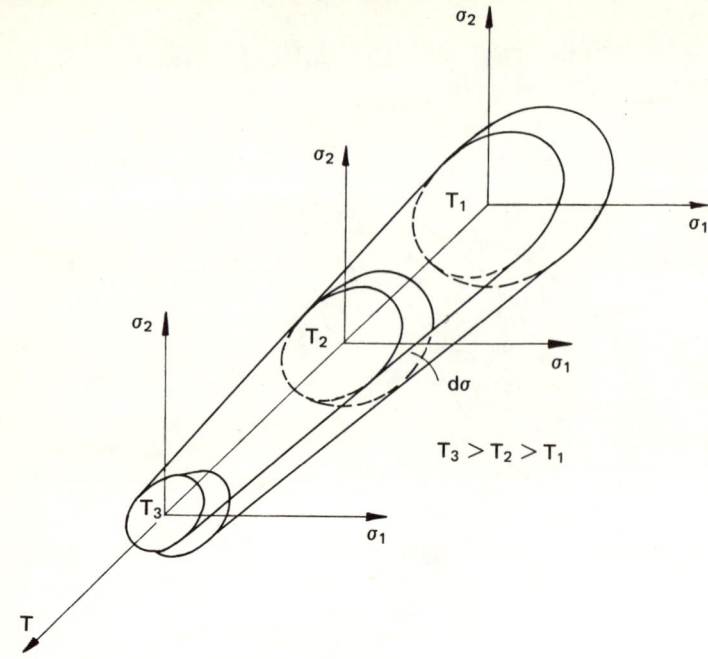

Figure 8.10. Yield surface as a function of stress and temperature.

The plastic potential function F can be expressed as
$$F = F(\sigma_{ij}, \alpha_{ij}, T, \dot{\varepsilon}). \tag{8.24}$$
The total differential of F is
$$\dot{F} = \frac{\partial F}{\partial \sigma_{ij}} \dot{\sigma}_{ij} + \frac{\partial F}{\partial \alpha_{ij}} \dot{\alpha}_{ij} + \frac{\partial F}{\partial T} \dot{T} + \frac{\partial F}{\partial \dot{\varepsilon}} (\dot{\varepsilon})^{\cdot} = 0. \tag{8.25}$$
The condition (8.25) is sometimes called the compatibility equation for the plastic yielding condition.

On the basis of (8.21) we can obtain the following relations
$$\frac{\partial F}{\partial \alpha_{ij}} = -\frac{\partial F}{\partial \sigma_{ij}}, \tag{8.26}$$
$$\frac{\partial \sigma_{ij}}{\partial \alpha_{ij}} = -1. \tag{8.27}$$
If one substitutes the above expressions into (8.25) the following relation is obtained
$$\frac{\partial F}{\partial \sigma_{ij}} (\dot{\sigma}_{ij} - \dot{\alpha}_{ij}) + \frac{\partial F}{\partial T} \dot{T} + \frac{\partial F}{\partial \dot{\varepsilon}} (\dot{\varepsilon})^{\cdot} = 0. \tag{8.28}$$

or

$$\frac{\partial F}{\partial \sigma_{ij}} \dot{\alpha}_{ij} = \frac{\partial F}{\partial \sigma_{ij}} \dot{\sigma}_{ij} + \frac{\partial F}{\partial T} \dot{T} + \frac{\partial F}{\partial \dot{\bar{\varepsilon}}} (\dot{\bar{\varepsilon}})^{\cdot}. \tag{8.29}$$

Further substitution of (8.21) into the above expression yields

$$d\mu = \frac{\left(\dfrac{\partial F}{\partial \sigma_{ij}} \dot{\sigma}_{ij} + \dfrac{\partial F}{\partial T} \dot{T} + \dfrac{\partial F}{\partial \dot{\bar{\varepsilon}}} (\dot{\bar{\varepsilon}})^{\cdot}\right)}{\left((\sigma_{kl} - \alpha_{kl}) \dfrac{\partial F}{\partial \sigma_{kl}}\right)}. \tag{8.30}$$

Considering the flow rule

$$\dot{\varepsilon}_{ij}^{p} = \lambda \frac{\partial F}{\partial \sigma_{ij}}, \tag{8.31}$$

the proportionality factor λ can be derived by combining (8.22) and the flow rule as follows

$$\lambda = \frac{1}{C(T, \dot{\bar{\varepsilon}})} \frac{\left(\dfrac{\partial F}{\partial \sigma_{ij}} \dot{\sigma}_{ij} + \dfrac{\partial F}{\partial T} \dot{T} + \dfrac{\partial F}{\partial \dot{\bar{\varepsilon}}} (\dot{\bar{\varepsilon}})^{\cdot}\right)}{\dfrac{\partial F}{\partial \sigma_{kl}} \dfrac{\partial F}{\partial \sigma_{kl}}}. \tag{8.32}$$

PART III

SMALL STRAIN THERMO-ELASTO-PLASTICITY

9 Equations for Thermo-Elasto-Plasticity

9.1 Isotropic Hardening

Assume the following decomposition for the rate of the strain tensor in a thermo-elasto-plastic process in a solid body

$$\dot{\varepsilon}_{ij} = \dot{\varepsilon}_{ij}^e + \dot{\varepsilon}_{ij}^T + \dot{\varepsilon}_{ij}^{e,T} + \dot{\varepsilon}_{ij}^{e,\dot{\varepsilon}} + \dot{\varepsilon}_{ij}^p \tag{9.1}$$

where $\dot{\varepsilon}_{ij}^p$ are the components of the plastic strain rate tensor. After rearranging (9.1) we can obtain the components of the elastic strain rate tensor

$$\dot{\varepsilon}_{ij}^e = \dot{\varepsilon}_{ij} - \dot{\varepsilon}_{ij}^T - \dot{\varepsilon}_{ij}^{e,T} - \dot{\varepsilon}_{ij}^{e,\dot{\varepsilon}} - \dot{\varepsilon}_{ij}^p. \tag{9.2}$$

Making use of Hooke's law the rates of change of the total stress are given as

$$\dot{\sigma}_{ij} = C_{ijkl}^e (\dot{\varepsilon}_{kl} - \dot{\varepsilon}_{kl}^T - \dot{\varepsilon}_{kl}^{e,T} - \dot{\varepsilon}_{kl}^{e,\dot{\varepsilon}} - \dot{\varepsilon}_{kl}^p). \tag{9.3}$$

Combining (8.20), (9.2) and (9.3) we get

$$\frac{\partial F}{\partial \sigma_{ij}} C_{ijkl}^e (\dot{\varepsilon}_{kl} - \dot{\varepsilon}_{kl}^T - \dot{\varepsilon}_{kl}^{e,T} - \dot{\varepsilon}_{kl}^{e,\dot{\varepsilon}} - \dot{\varepsilon}_{kl}^p) + \frac{\partial F}{\partial K} \frac{\partial K}{\partial \varepsilon_{ij}^p} \frac{\partial F}{\partial \sigma_{ij}} \lambda$$
$$+ \frac{\partial F}{\partial T} \dot{T} + \frac{\partial F}{\partial \dot{\varepsilon}} (\dot{\varepsilon})^{\cdot} = 0. \tag{9.4}$$

After some calculations we obtain the relation for the proportionality factor λ

$$\lambda = \frac{\left(\frac{\partial F}{\partial \sigma_{ij}} C_{ijkl}^e (\dot{\varepsilon}_{kl} - \dot{\varepsilon}_{kl}^T - \dot{\varepsilon}_{kl}^{e,T} - \dot{\varepsilon}_{kl}^{e,\dot{\varepsilon}}) + \frac{\partial F}{\partial T} \dot{T} + \frac{\partial F}{\partial \dot{\varepsilon}} (\dot{\varepsilon})^{\cdot} \right)}{\left(\frac{\partial F}{\partial \sigma_{pq}} C_{pqrs}^e \frac{\partial F}{\partial \sigma_{rs}} - \frac{\partial F}{\partial K} \frac{\partial K}{\partial \varepsilon_{pq}^p} \frac{\partial F}{\partial \sigma_{pq}} \right)}. \tag{9.5}$$

If we introduce the notations

$$S = \frac{\partial F}{\partial \sigma_{pq}} C_{pqrs}^e \frac{\partial F}{\partial \sigma_{rs}} - \frac{\partial F}{\partial K} \frac{\partial K}{\partial \varepsilon_{pq}^p} \frac{\partial F}{\partial \sigma_{pq}} \tag{9.6}$$

and

$$\dot{\varepsilon}_{ij}^T = \alpha_{ij}\dot{T},$$

$$D_{ij}^T = \frac{\partial E_{ijkl}}{\partial T}\sigma_{kl}, \tag{9.7}$$

$$D_{ij}^{e,\dot{\bar{\varepsilon}}} = \frac{\partial E_{ijkl}}{\partial \dot{\bar{\varepsilon}}}\sigma_{kl}, \tag{9.8}$$

then from (9.5) and (9.6) we get

$$\lambda = \frac{1}{S}\left(\frac{\partial F}{\partial \sigma_{ij}}C_{ijkl}^e(\dot{\varepsilon}_{kl} - \alpha_{kl}\dot{T} - D_{kl}^T\dot{T} - D_{kl}^{e,\dot{\bar{\varepsilon}}}(\dot{\bar{\varepsilon}})^{\cdot}) + \frac{\partial F}{\partial T}\dot{T} + \frac{\partial F}{\partial \dot{\bar{\varepsilon}}}(\dot{\bar{\varepsilon}})^{\cdot}\right). \tag{9.9}$$

Next combining (9.3) and (9.9) we have

$$\dot{\sigma}_{ij} = C_{ijkl}^e\dot{\varepsilon}_{kl} - C_{ikjl}^e(\alpha_{kl}\dot{T} + D_{kl}^T\dot{T} + D_{kl}^{e,\dot{\bar{\varepsilon}}}(\dot{\bar{\varepsilon}})^{\cdot}) - \frac{C_{ijvw}^e}{S}\frac{\partial F}{\partial \sigma_{vw}}$$

$$\times \left(\frac{\partial F}{\partial \sigma_{tu}}C_{tukl}^e(\dot{\varepsilon}_{kl} - \alpha_{kl}\dot{T} - D_{kl}^T\dot{T} - D_{kl}^{e,\dot{\bar{\varepsilon}}}(\dot{\bar{\varepsilon}})^{\cdot}) + \frac{\partial F}{\partial T}\dot{T} + \frac{\partial F}{\partial \dot{\bar{\varepsilon}}}(\dot{\bar{\varepsilon}})^{\cdot}\right). \tag{9.10}$$

After rearranging the terms

$$\dot{\sigma}_{ij} = C_{ijkl}^e\dot{\varepsilon}_{kl} - C_{ijkl}^e(\alpha_{kl}\dot{T} + D_{kl}^T\dot{T} + D_{kl}^{e,\dot{\bar{\varepsilon}}}(\dot{\bar{\varepsilon}})^{\cdot}) - \frac{1}{S}C_{ijvw}^e\frac{\partial F}{\partial \sigma_{vw}}\frac{\partial F}{\partial \sigma_{tu}}C_{tukl}^e\dot{\varepsilon}_{kl}$$

$$+ \frac{1}{S}C_{ijvw}^e\frac{\partial F}{\partial \sigma_{vw}}\frac{\partial F}{\partial \sigma_{tu}}C_{tukl}^e(\alpha_{kl}\dot{T} + D_{kl}^T\dot{T} + D_{kl}^{e,\dot{\bar{\varepsilon}}}(\dot{\bar{\varepsilon}})^{\cdot}) \tag{9.11}$$

$$- \frac{1}{S}C_{ijkl}^e\frac{\partial F}{\partial \sigma_{kl}}\left(\frac{\partial F}{\partial T}\dot{T} + \frac{\partial F}{\partial \dot{\bar{\varepsilon}}}(\dot{\bar{\varepsilon}})^{\cdot}\right).$$

We can define the so-called plasticity tensor

$$C_{ijkl}^p = \frac{1}{S}C_{ijvw}^e\frac{\partial F}{\partial \sigma_{vw}}\frac{\partial F}{\partial \sigma_{tu}}C_{tukl}^e = \frac{1}{S}C_{ijvw}^eS_{vw}S_{tu}C_{tukl}^e. \tag{9.12}$$

By substituting (9.12) into (9.11) we get

$$\dot{\sigma}_{ij} = C_{ijkl}^e\dot{\varepsilon}_{kl} - C_{ijkl}^e(\alpha_{kl}\dot{T} + D_{kl}^T\dot{T} + D_{kl}^{e,\dot{\bar{\varepsilon}}}(\dot{\bar{\varepsilon}})^{\cdot}) - C_{ijkl}^p\dot{\varepsilon}_{kl} + C_{ijkl}^p(\alpha_{kl}\dot{T}$$

$$+ D_{kl}^T\dot{T} + D_{kl}^{e,\dot{\bar{\varepsilon}}}(\dot{\bar{\varepsilon}})^{\cdot}) - \frac{1}{S}C_{ijkl}^e\frac{\partial F}{\partial \sigma_{kl}}\left(\frac{\partial F}{\partial T}\dot{T} + \frac{\partial F}{\partial \dot{\bar{\varepsilon}}}(\dot{\bar{\varepsilon}})^{\cdot}\right). \tag{9.13}$$

Defining the so-called elasto-plasticity tensor

$$C_{ijkl}^{ep} = C_{ijkl}^e - C_{ijkl}^p, \tag{9.14}$$

the thermo-elasto-plastic constitutive equation can be expressed as

$$\dot{\sigma}_{ij} = C_{ijkl}^{ep}\dot{\varepsilon}_{kl} - C_{ijkl}^{ep}(\alpha_{kl}\dot{T} + D_{kl}^T\dot{T} + D_{kl}^{e,\dot{\bar{\varepsilon}}}(\dot{\bar{\varepsilon}})^{\cdot})$$

$$- \frac{C_{ijkl}^eS_{kl}}{S}\left(\frac{\partial F}{\partial T}\dot{T} + \frac{\partial F}{\partial \dot{\bar{\varepsilon}}}(\dot{\bar{\varepsilon}})^{\cdot}\right). \tag{9.15}$$

One may find that during plastic deformation of a solid

$$C^e_{ijkl}s_{kl} = 2Gs_{ij} \tag{9.16}$$

due to the fact that the first deviatoric strain invariant vanishes. It leads to the following relation on C^p_{ijkl}

$$C^e_{ijvw}s_{vw}s_{tu}C^e_{tukl} = 2Gs_{ij}s_{kl}2G = 4G^2 s_{ij}s_{kl}. \tag{9.17}$$

Now we evaluate the S given by (9.6). Since $C^e_{pqrs}s_{rs} = 2Gs_{pq}$ the first term in Eq. (9.6) can be expressed as

$$\frac{\partial F}{\partial \sigma_{pq}} C^e_{pqrs} \frac{\partial F}{\partial \sigma_{rs}} = s_{pq}C^e_{pqrs}s_{rs} = s_{pq}2Gs_{pq} = 2G\frac{2\bar{\sigma}^2}{3} = \tfrac{4}{3}G\bar{\sigma}^2, \tag{9.18}$$

where

$$\tfrac{1}{3}\bar{\sigma}^2 = \tfrac{1}{2}s_{pq}s_{pq}. \tag{9.19}$$

The work-hardening parameter K of the material is taken to be represented by the amount of plastic work done during the plastic deformation

$$dK = \sigma_{ij}d\varepsilon^p_{ij}. \tag{9.20}$$

Substituting (8.9) and (8.10) into the above gives

$$dK = \lambda \sigma_{ij} \frac{\partial F}{\partial \sigma_{ij}}. \tag{9.21}$$

From (9.20) we get

$$\sigma_{ij} = \frac{dK}{d\varepsilon^p_{ij}}. \tag{9.22}$$

During plastic deformation of work-hardening materials, the yield strength increases with load. The definitions of the HM plastic potential function can be used to derive the following useful relation

$$\frac{\partial F}{\partial K} = \frac{\partial}{\partial K}\left(J_{2s} - \frac{1}{3}\bar{\sigma}^2\right) = -\frac{2}{3}\bar{\sigma}\frac{\partial \bar{\sigma}}{\partial K}. \tag{9.23}$$

If we refer to a typical flow curve we have

$$dK = \bar{\sigma}d\bar{\varepsilon}^p, \tag{9.24}$$

or

$$\frac{d\bar{\varepsilon}^p}{dK} = \frac{1}{\bar{\sigma}}.$$

Since

$$\frac{\partial \bar{\sigma}}{\partial K} = \frac{\partial \bar{\sigma}}{\partial \bar{\varepsilon}^p}\frac{d\bar{\varepsilon}^p}{dK} = H'\cdot\frac{1}{\bar{\sigma}} = \frac{H'}{\bar{\sigma}} \tag{9.25}$$

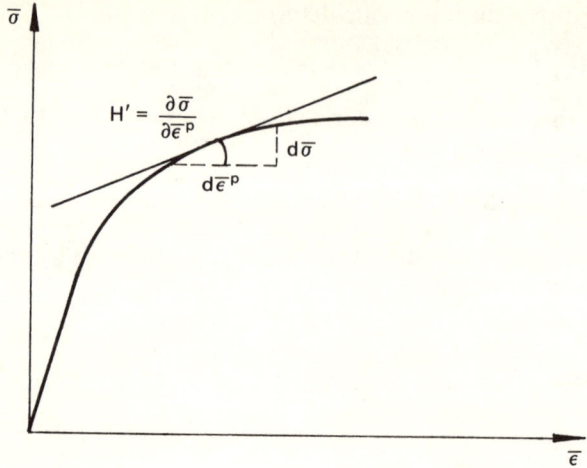

Figure 9.1. Slope H' as a plastic modulus of the material during plastic deformation.

where H' is the plastic modulus of the material in a multiaxial stress state (see Fig. 9.1)

$$H' = \frac{\partial \bar{\sigma}}{\partial \bar{\varepsilon}^p}, \qquad (9.26)$$

Eq. (9.23) will have the form

$$\frac{\partial F}{\partial K} = -\frac{2}{3}\bar{\sigma}\left(\frac{H'}{\bar{\sigma}}\right) = -\frac{2H'}{3} \qquad (9.27)$$

which leads to the following relation

$$\frac{\partial F}{\partial K}\frac{\partial K}{\partial \varepsilon^p_{pq}}\frac{\partial F}{\partial \sigma_{pq}} = -\frac{2H'}{3}\sigma_{pq}s_{sq} = -\frac{2}{3}H'\frac{2}{3}\bar{\sigma}^2 = -\frac{4}{9}\bar{\sigma}^2 H'. \qquad (9.28)$$

Then S can be expressed in terms of the material properties and the state of stress as follows

$$S = \frac{4}{3}G\bar{\sigma}^2 - \left(-\frac{4}{9}\bar{\sigma}^2 H'\right) = -\frac{4}{9}\bar{\sigma}^2(3G + H'). \qquad (9.29)$$

9.2 Kinematic Hardening

In kinematic hardening theory we define the so-called translated stress tensor

$$\sigma^*_{ij} = \sigma_{ij} - \alpha_{ij}, \qquad (9.30)$$

and translated stress deviators

$$s^*_{ij} = \sigma^*_{ij} - \tfrac{1}{3}\sigma^*_{kk}\delta_{ij}. \qquad (9.31)$$

EQUATIONS FOR THERMO-ELASTO-PLASTICITY

The yield function becomes

$$F = \tfrac{2}{3} s_{ij}^* s_{ij}^*, \tag{9.32}$$

which leads to the expression

$$\frac{\partial F}{\partial \sigma_{ij}} = 3 s_{ij}^* \frac{\partial s_{ij}^*}{\partial \sigma_{ij}} = 6 s_{ij}^*. \tag{9.33}$$

By substituting (9.33) into (8.30), (8.31) and (8.32) we get

$$d\mu = \frac{6 s_{ij}^* \dot{\sigma}_{ij} + \dfrac{\partial F}{\partial T} \dot{T} + \dfrac{\partial F}{\partial \dot{\bar{\varepsilon}}} (\dot{\bar{\varepsilon}})^{\cdot}}{6 \sigma_{pq}^* s_{pq}^*}, \tag{9.34}$$

$$\lambda = \frac{1}{C(T, \dot{\bar{\varepsilon}})} \frac{6 s_{ij}^* \dot{\sigma}_{ij} + \dfrac{\partial F}{\partial T} \dot{T} + \dfrac{\partial F}{\partial \dot{\bar{\varepsilon}}} (\dot{\bar{\varepsilon}})^{\cdot}}{36 s_{pq}^* s_{pq}^*}, \tag{9.35}$$

$$\dot{\varepsilon}_{ij}^p = 6 s_{ij}^* \lambda. \tag{9.36}$$

Assuming

$$\dot{\varepsilon}_{ij} = \dot{\varepsilon}_{ij}^e + \dot{\varepsilon}_{ij}^T + \dot{\varepsilon}_{ij}^{e;T} + \dot{\varepsilon}_{ij}^{e;\dot{\bar{\varepsilon}}} + \dot{\varepsilon}_{ij}^p, \tag{9.37}$$

and using Hooke's law one obtains

$$\dot{\sigma}_{ij} = C_{ijkl}^e (\dot{\varepsilon}_{kl} - \dot{\varepsilon}_{kl}^T - \dot{\varepsilon}_{kl}^{e;T} - \dot{\varepsilon}_{kl}^{e;\dot{\bar{\varepsilon}}} - \dot{\varepsilon}_{kl}^p), \tag{9.38}$$

or

$$\dot{\sigma}_{ij} = C_{ijkl}^e \dot{\varepsilon}_{kl} - C_{ijkl}^e (\alpha_{kl} \dot{T} + D_{kl}^T \dot{T} + D_{kl}^{\dot{\bar{\varepsilon}}} (\dot{\bar{\varepsilon}})^{\cdot} + 6 s_{kl}^* \lambda) \tag{9.39}$$

where

$$D_{kl}^T = \frac{\partial E_{klmn}}{\partial T} \sigma_{mn}, \tag{9.40}$$

$$D_{kl}^{\dot{\bar{\varepsilon}}} = \frac{\partial E_{klmn}}{\partial \dot{\bar{\varepsilon}}} \sigma_{mn}. \tag{9.41}$$

Substituting (8.22) and (9.33) into (8.31) gives

$$d\alpha_{ij} = 6 C(T, \dot{\bar{\varepsilon}}) s_{ij}^* \lambda. \tag{9.42}$$

The above expression, after substituting into (8.25), gives

$$\dot{F} = (\dot{\sigma}_{ij} - 6 C(T, \dot{\bar{\varepsilon}}) s_{ij}^* \lambda) \frac{\partial F}{\partial \sigma_{ij}} + \frac{\partial F}{\partial T} \dot{T} + \frac{\partial F}{\partial \dot{\bar{\varepsilon}}} (\dot{\bar{\varepsilon}})^{\cdot} = 0. \tag{9.43}$$

By (9.43) and (9.35) we get

$$\lambda = \frac{\dfrac{\partial F}{\partial \sigma_{ij}} (C_{ijkl}^e \dot{\varepsilon}_{kl} - C_{ijkl}^e (\alpha_{kl} \dot{T} + D_{kl}^T \dot{T} + D_{kl}^{\dot{\bar{\varepsilon}}} (\dot{\bar{\varepsilon}})^{\cdot})) + \dfrac{\partial F}{\partial T} \dot{T} + \dfrac{\partial F}{\partial \dot{\bar{\varepsilon}}} (\dot{\bar{\varepsilon}})^{\cdot}}{6 (C(T, \dot{\bar{\varepsilon}}) s_{pq}^* + C_{pqrs}^e s_{rs}^*) \dfrac{\partial F}{\partial \sigma_{pq}}}. \tag{9.44}$$

Substituting (9.44) into (9.39) we get

$$\dot{\sigma}_{ij} = C^e_{ijkl}\dot{\varepsilon}_{kl} - C^e_{ijkl}(\alpha_{kl}\dot{T} + D^T_{kl}\dot{T} + D^{\dot{\varepsilon}}_{kl}(\dot{\varepsilon}))$$

$$- C^e_{ijvw}s^*_{vw} \frac{\frac{\partial F}{\partial \sigma_{tu}}(C^e_{tukl}\dot{\varepsilon}_{kl} - C^e_{tukl}(\alpha_{kl}\dot{T} + D^T_{kl}\dot{T} + D^{\dot{\varepsilon}}_{kl}(\dot{\varepsilon})') + \frac{\partial F}{\partial T}\dot{T} + \frac{\partial F}{\partial \dot{\varepsilon}}(\dot{\varepsilon})'}{(C(T,\dot{\varepsilon})s^*_{pq} + C^e_{pqrs}s^*_{rs}) \frac{\partial F}{\partial \sigma_{pq}}}$$

(9.45)

Let

$$S = (C(T,\dot{\varepsilon})s^*_{pq} + C^e_{pqrs}s^*_{rs}) \frac{\partial F}{\partial \sigma_{pq}} = 6s^*_{pq}s^*_{pq}C(T,\dot{\varepsilon}) + 6s^*_{pq}C^e_{pqrs}s^*_{rs}.$$

(9.46)

With the aid of (9.46) the relation (9.45) becomes

$$\dot{\sigma}_{ij} = C^e_{ijkl}\dot{\varepsilon}_{kl} - C^e_{ijkl}(\alpha_{kl}\dot{T} + D^T_{kl}\dot{T} + D^{\dot{\varepsilon}}_{kl}(\dot{\varepsilon})')$$

$$- C^e_{ijvw}s^*_{vw} \frac{\partial F}{\partial \sigma_{tu}} \frac{1}{S}(C^e_{tukl}\dot{\varepsilon}_{kl} - C^e_{tukl}(\alpha_{kl}\dot{T} + D^T_{kl}\dot{T} + D^{\dot{\varepsilon}}_{kl}(\dot{\varepsilon})'))$$

$$- \frac{1}{S}C^e_{ijvw}s^*_{vw}\left(\frac{\partial F}{\partial T}\dot{T} + \frac{\partial F}{\partial \dot{\varepsilon}}(\dot{\varepsilon})'\right)$$

$$= \left(C^e_{ijkl} - \frac{1}{S}\left(C^e_{ijvw}s^*_{vw}\frac{\partial F}{\partial \sigma_{tu}}C^e_{tukl}\right)\right)\dot{\varepsilon}_{kl}$$

$$- \left(C^e_{ijkl} - \frac{1}{S}\left(C^e_{ijvw}s^*_{vw}\frac{\partial F}{\partial \sigma_{tu}}C^e_{tukl}\right)\right)(\alpha_{kl}\dot{T} + D^T_{kl}\dot{T} + D^{\dot{\varepsilon}}_{kl}(\dot{\varepsilon}))$$

$$- \frac{1}{S}C^e_{ijvw}s^*_{vw}\left(\frac{\partial F}{\partial T}\dot{T} + \frac{\partial F}{\partial \dot{\varepsilon}}(\dot{\varepsilon})'\right).$$

(9.47)

Assuming that

$$\frac{1}{S}\left(C^e_{ijvw}s^*_{vw}\frac{\partial F}{\partial \sigma_{tu}}C^e_{tukl}\right) = \frac{1}{S}(6C^e_{ijvw}s^*_{vw}s^*_{tu}C^e_{tukl}) = C^p_{ijkl}$$

(9.48)

and

$$C^{ep}_{ijkl} = C^e_{ijkl} - C^p_{ijkl},$$

(9.49)

the constitutive equation for kinematic hardening material subject to thermo-elasto-plastic deformation takes the form

$$\dot{\sigma}_{ij} = C^{ep}_{ijkl}\dot{\varepsilon}_{kl} - C^{ep}_{ijkl}(\alpha_{kl}\dot{T} + D^T_{kl}\dot{T} + D^{\dot{\varepsilon}}_{kl}(\dot{\varepsilon})')$$

$$- \frac{1}{S}C^e_{ijkl}s^*_{kl}\left(\frac{\partial F}{\partial T} + \frac{\partial F}{\partial \dot{\varepsilon}}(\dot{\varepsilon})\right).$$

(9.50)

9.3 Elasto-Visco-Plasticity

Small strain thermo-elasto-visco-plasticity theory is used in the analysis of welding and casting processes. The additive decomposition for the rate of the strain tensor in a thermo-elasto-visco-plastic deformation in a solid body has the form

$$\dot{\varepsilon}_{ij} = \dot{\varepsilon}^e_{ij} + \dot{\varepsilon}^T_{ij} + \dot{\varepsilon}^{e,T}_{ij} + \dot{\varepsilon}^{e,\dot{\varepsilon}}_{ij} + \dot{\varepsilon}^{vp}_{ij}. \tag{9.51}$$

where $\dot{\varepsilon}^{vp}_{ij}$ are the components of the visco-plastic strain rate tensor.

The visco-plastic strain rate tensor components are assumed to be given by a flow rule of the form

$$\dot{\varepsilon}^{vp}_{ij} = \gamma \langle \varphi(h) \rangle_{(F)} \frac{\partial F}{\partial \sigma_{ij}} \tag{9.52}$$

where $\gamma = \gamma(T)$ is a temperature-dependent material viscosity coefficient, φ is a discontinuous function which ensures that no visco-plastic flow occurs below the yield condition $F = 0$

$$\langle \varphi(h) \rangle_{(F)} = \begin{cases} 0 & \text{for } F \leq 0, \\ \varphi(h) & \text{for } F > 0, \end{cases} \tag{9.53}$$

and h is a scalar measure of the excess stress to be defined. The function $\varphi(h)$ is determined empirically.

Making use of Hooke's law the changes in total stress are equal to

$$\dot{\sigma}_{ij} = C^e_{ijkl} \left(\dot{\varepsilon}^e_{kl} - \dot{\varepsilon}^T_{kl} - \dot{\varepsilon}^{e,T}_{kl} - \dot{\varepsilon}^{e,\dot{\varepsilon}}_{kl} - \gamma \langle \varphi(h) \rangle \frac{\partial F}{\partial \sigma_{ij}} \right). \tag{9.54}$$

There exist several stress–strain relations which are used to describe visco-plastic material behaviour. Some of them are given below

$$\sigma = \sigma_0 + \sigma_1 \ln \frac{\dot{\varepsilon}}{\dot{\varepsilon}_0} \qquad \text{–Ludwik (1909)} \tag{9.55}$$

$$\sigma = f(\varepsilon) + a \ln(1 + b\dot{\varepsilon}) \qquad \text{–Malvern (1951a)} \tag{9.56}$$

$$\dot{\varepsilon} = \dot{\varepsilon}^e + \dot{\varepsilon}^{vp} = \frac{\dot{\sigma}}{E} + F(\sigma - f(\varepsilon)) \qquad \text{–Malvern (1951b)} \tag{9.57}$$

$$\dot{\varepsilon}^{vp} = k(\sigma - \sigma_0) \qquad \text{–Sokołowski (1950)} \tag{9.58}$$

$$\dot{\varepsilon}^{vp} = D \left(\frac{\sigma}{\sigma_0} - 1 \right)^p \sin \sigma \qquad \text{–Cowper and Symonds (1957)} \tag{9.59}$$

$$\dot{\varepsilon}^{vp} = \left[D \left(\frac{\sigma}{\sigma_0} - 1 + H \right)^p + H^p \right] \sin \sigma \qquad \text{–Clark and Duwez (1950)} \tag{9.60}$$

$$\dot{\varepsilon}^{vp} = \sum_{j=1}^{n} D_j \left(\frac{\sigma}{\sigma_0} - 1 \right)^j \qquad \text{–Perzyna (1966)} \tag{9.61}$$

where $a, b, D, D_j, H, k, p, \sigma_0, \sigma_1, \dot{\varepsilon}_0$ are material constants.

10 Finite-Element Solution

10.1 Finite-Element Solution of Heat Flow Equations

10.1.1 Weighted Residual Method

Heat transfer problems can be formulated either by a given differential equation with boundary conditions or by a given functional equivalent to the differential equation (Kleiber and Służalec 1983a, b, 1984; Służalec 1987a, b, 1988a). The weighted residual method is an approximate solution method for differential equations. This method widens the range of problems amenable to solution since it does not require a variational formulation of the problem. Consider the equation

$$\nabla \cdot (k\nabla T) + q_v = \rho c_p \frac{\partial T}{\partial t} \tag{10.1}$$

with boundary conditions

$$B_1(T) = T - T_w = 0 \qquad \text{on } S_1, \tag{10.2}$$

$$B_2(T) = k\frac{\partial T}{\partial n} + q_w = 0 \qquad \text{on } S_2, \tag{10.3}$$

$$B_3(T) = k\frac{\partial T}{\partial n} + \alpha(T - T_f) = 0 \qquad \text{on } S_3, \tag{10.4}$$

where T_w is the temperature of the body surface and T_f is the fluid temperature. We look for a solution of the problem (10.1)–(10.4) in the class of functions $\tilde{T}$, fulfilling the boundary conditions (10.2)–(10.4). These functions in general do not fulfil the differential equation (10.1). Substituting one of these functions into Eq. (10.1) we get

$$\nabla \cdot (k\nabla T) + q_v - \rho c_p \frac{\partial T}{\partial t} = E \neq 0. \tag{10.5}$$

The best approximation for the solution will be a function T which minimizes the residuum E. The simplest method of obtaining the solution is to make use of the fact that if E is identically equal to zero, then

$$\int_V wE\,dV = 0, \tag{10.6}$$

where V is the volume of the body under consideration and w is an arbitrary function.

A similar procedure is used for boundary conditions. We get

$$\int_V w \left[\nabla(k\nabla T) + q_v - \rho c_p \frac{\partial T}{\partial t} \right] dV = 0, \tag{10.7}$$

and

$$\sum_{i=1}^{3} \int_{S_i} w_i B_i(T) dS_i = 0. \tag{10.8}$$

The above conditions may be replaced by the single condition

$$\int_V w_0 \left[\nabla(k\nabla T) + q_v - \rho c_p \frac{\partial T}{\partial t} \right] dV + \sum_{i=1}^{3} \int_{S_i} w_i B_i(T) dS_i = 0. \tag{10.9}$$

Since we are dealing with a second-order partial differential operator the function T must be continuous with continuous first-order derivatives (i.e. of class C^1). If we use the divergence theorem to lessen the order of differentiation in (10.9) then we may simply require the temperature to be of class C^0. Then we can express it in terms of shape functions, for which this condition is fulfilled. Using the divergence theorem, the requirements as to the function w increase; it now has to be a condition function. Making use of the divergence theorem for Eq. (10.9) we get

$$\int_V \nabla w(k\nabla T) dV - \int_V wq_v dV - \int_V w\rho c_p \frac{\partial T}{\partial t} dV - \int_{S_1} wk \frac{\partial T}{\partial n} dS$$

$$- \int_{S_2} w_2 \left[k \frac{\partial T}{\partial n} + q_w \right] dS - \int_{S_3} w_3 \left[k \frac{\partial T}{\partial n} + \alpha(T - T_f) \right] dS = 0, \tag{10.10}$$

where $S = S_1 \cup S_2 \cup S_3$ is the total surface of the body.

Since functions w_i are arbitrary continuous functions, we can put

$$w_2 = w_3 = -w. \tag{10.11}$$

Finally we get

$$\int_V \nabla w k \nabla T dV - \int_V wq_v dV - \int_V w\rho c_p \frac{\partial T}{\partial t} dV + \int_{S_2} wq_w dS$$

$$+ \int_{S_3} w\alpha(T - T_f) dS - \int_{S_1} wk \frac{\partial T}{\partial n} dS = 0. \tag{10.12}$$

Assume that the temperature can be approximated by the expression

$$T = \sum_{i=1}^{I} H_i T_i, \tag{10.13}$$

where H_i are shape functions, T_i are nodal temperatures and I stands for the number of nodes. The expression (10.13) is sometimes shown in matrix notation as

$$T = H^T T$$

where H is the vector of shape functions and T is the vector of nodal temperatures. Functions w in Eq. (10.12) can be taken as

$$w = H_i. \tag{10.14}$$

This assumption reduces the weighted residual method to the classical Galerkin method. The particular advantage of such an approach is that for heat exchange problems in non-moving bodies the matrix of the system is symmetric and banded.

Substitute in Eq. (10.12) the functions

$$w = H_i, \qquad i = 1, 2, \ldots, I \tag{10.15}$$

By substitution of temperature T in the form of (10.13) we get a system of I equations with I unknowns, T_i:

$$\mathbf{K}T + \mathbf{C}\dot{T} = \mathbf{F}, \tag{10.16}$$

where

$$K_{ij} = \int_V \nabla H_i k \nabla H_j \, dV + \int_{S_3} H_i \alpha H_j \, dS - \int_{S_1} H_i k \frac{\partial H_j}{\partial n} \, dS, \tag{10.17}$$

$$C_{ij} = \int_V H_i c_p \rho H_j \, dV, \tag{10.18}$$

$$F_i = \int_V H_i q_v \, dV - \int_{S_2} H_i q_w \, dS + \int_{S_3} H_i \alpha T_f \, dS. \tag{10.19}$$

As we see, the matrix $\mathbf{K}$ is symmetric. Integrals in expressions (10.17) and (10.18) are different from zero only in the sub-domains where $H_i \neq 0$, e.g. in the elements which possess the ith node.

10.1.2 Variational Formulation

The heat exchange problem is uniquely defined by the partial differential equation with boundary conditions. It is however possible to formulate the problem using the extreme variational principle. According to Euler's theorem the solution of the partial differential equation of heat transfer

$$k_{11} \frac{\partial^2 T}{\partial x^2} + k_{22} \frac{\partial^2 T}{\partial y^2} + k_{33} \frac{\partial^2 T}{\partial z^2} + q_v = \rho c_p \frac{\partial T}{\partial t} \tag{10.20}$$

may be found by minimizing the functional

$$\Pi = \frac{1}{2} \int_V \left\{ \left[k_{11} \left(\frac{\partial T}{\partial x} \right)^2 + k_{22} \left(\frac{\partial T}{\partial y} \right)^2 + k_{33} \left(\frac{\partial T}{\partial z} \right)^2 \right] \right. \\ \left. - 2 q_v T + \rho c T^2 + 2 \rho c T_0 T \right\} dV \tag{10.21}$$

with assumptions that the function T fulfils the given boundary conditions

$$T = T_w \quad \text{on } S_1, \tag{10.22}$$

$$k\frac{\partial T}{\partial n} + q_w = 0 \quad \text{on } S_2, \tag{10.23}$$

$$k\frac{\partial T}{\partial n} + \alpha(T - T_f) = 0 \quad \text{on } S_3, \tag{10.24}$$

where T_0 is the initial spatial temperature distribution.

The condition (10.22) can be easily fulfilled. However the remaining conditions create significant difficulties. Therefore one adds to the functional (10.21) the surface integral, which in the process of minimization leads to (10.23) or (10.24). In general it has the form

$$\frac{1}{2}\left\{\int_{S_2} 2q_w T dS + \int_{S_3} \alpha(T^2 - 2TT_f)\, dS\right\} \tag{10.25}$$

or

$$\bar{\Pi} = \Pi + \frac{1}{2}\left\{\int_{S_2} 2q_w T dS + \int_{S_3} \alpha(T^2 - 2TT_f)\, dS\right\} \tag{10.26}$$

and at the same time the function T has to fulfil the condition (10.22).

The functional in the form (10.26) can be expressed as the sum of functionals determined for each differential element

$$\Pi = \sum_n {}^n\bar{\Pi} \tag{10.27}$$

It reaches a minimum, if all components of the sum are minimal. Assume that the temperature T is expressed by shape functions and nodal temperatures

$$T = \sum_i H_i T_i. \tag{10.28}$$

Since all shape functions are continuous functions, derivatives appearing in the functional (10.26) are determined. In order to minimize this functional with respect to parameters T_i, we should solve the following system of equations:

$$\frac{\partial \bar{\Pi}}{\partial T_i} = 0 \quad \text{for } i = 1, 2, \ldots, I. \tag{10.29}$$

Finally we get

$$\mathbf{KT} + \mathbf{C}\dot{\mathbf{T}} = \mathbf{F}, \tag{10.30}$$

where

$$K_{ij} = \int_V \nabla H_i k \nabla H_j dV + \int_{S_3} H_i \alpha H_j dS - \int_{S_1} H_i k \frac{\partial H_j}{\partial n} dS, \tag{10.31}$$

$$C_{ij} = \int_V H_i c_p \rho H_j dV, \tag{10.32}$$

$$F_i = \int_V H_i q_v dV - \int_{S_2} H_i q_w dS + \int_{S_3} H_i \alpha T_f dS \tag{10.33}$$

and at the same time the surface integrals appear only in the cases where nodes are placed on the body boundary.

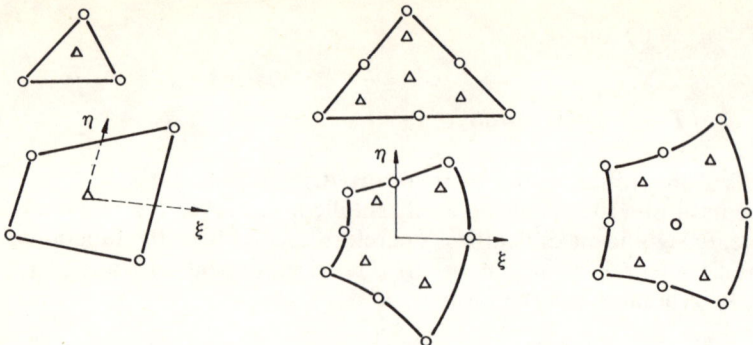

Figure 10.1. Optimal sampling points for the first derivatives in two-dimensional problems.

Consideration of the mathematical requirement in the general approximation leads to the conclusion that the error occurring in a finite-element approximation of T will be of order

$$O(d^{p+1}) \tag{10.34}$$

where d is the size of the element, and p is the order of polynomials used in the finite element expansions. This follows from consideration of the properties of a local Taylor series expansion (Służalec 1988e).

Similarly it is evident that the mth derivatives of T will be approximated to the order

$$O(d^{p-m+1}). \tag{10.35}$$

Thus, for instance, the first derivative will have an approximation of the order $O(d^p)$ only, and in many cases such a low order of approximation is undesirable, as the first derivative (corresponding to flux in thermal calculations) is of primary interest. In such cases fortunately we have recourse to a theorem which shows that for all elements (in a problem which is self-adjoint) there exist points at which the first derivatives have one order of higher approximation. Such points often coincide with certain Gaussian quadrature points which are used for numerical integration of the element integrals. In Fig. 10.1 we show some optimal sampling points for several commonly used finite elements in two-dimensional regions.

10.1.3 Time Integration Schemes for Nonlinear Heat Conduction

In this section we discuss various algorithms that have been proposed for numerical integration after finite-element discretization. The algorithms reviewed comprise one-step and two-step schemes. The study is an attempt to define criteria for an optimum choice among such algorithms, where emphasis is given to the accuracy achievable.

Consider the non-linear transient heat transfer matrix equations (10.16)–(10.19)

$$\mathbf{K}(T)T(t) + \mathbf{C}(T)\dot{T}(t) = F(T, t), \tag{10.36}$$
$$T(0) = T_0,$$

where $T(t)$ is the vector of the nodal temperatures (with a superscript dot denoting its time derivative), $\mathbf{K}$ is the symmetrical and positive semi-definite conductivity matrix, $\mathbf{C}$ is the symmetrical and positive definite capacity matrix and F is a vector of thermal loads corresponding to T; T_0 is a vector of given initial temperatures. In this section we shall consider and compare some of the methods currently used and proposed for the second stage of this solution procedure, i.e. the time integration of system (10.36). Thus we leave completely apart the problem of iterative solution of the resulting set of non-linear algebraic equations.

In order to undertake such a review and to facilitate the comparison with computational techniques applied to typical first-order differential equations, we rewrite system (10.36) in the form

$$\dot{T}(t) = -\mathbf{A}(T)T(t) + D(T, t) = G(T, t), \tag{10.37}$$
$$T(0) = T_0,$$

where $\mathbf{A} = \mathbf{C}^{-1}\mathbf{K}$ has real positive eigenvalues and $D = \mathbf{C}^{-1}F$ is a new forcing vector. Such a system is a stiff one, i.e. composed of exponentially decaying components with widely spread time constants which, in addition, change continuously in the non-linear case.

In current heat conduction applications one deals with situations where the slowly decaying components dominate the response; hence attention will be focused on time integration techniques that treat adequately the long-term components of the response while retaining numerical stability with respect to the fast-varying excitations. The simplest kind of time integration schemes are one-step schemes in which a two-level difference approximation is chosen for $\dot{T}$ and a linear variation of G is assumed over the interval $[t_n, t_{n+1}]$, thus yielding

$$T_{n+1} - T_n = h[(1-\theta)G_n + \theta G_{n+1}] \quad 0 \le \theta \le 1, \tag{10.38}$$

where h is the time step and G_n stands for $G(T_n, nh)$.

These methods will be referred to as generalized trapezoidal schemes. A slightly different form of one-step scheme is obtained from the same two-level difference approximation for $\dot{T}$ and from an assumed linear behaviour of T over $[t_n, t_{n+1}]$

$$T_{n+1} - T_n = hG_\theta = hG(T_\theta, t_\theta) \tag{10.39}$$

in which

$$T_\theta = (1-\theta)T_n + \theta T_{n+1},$$
$$\quad\quad\quad\quad\quad\quad\quad\quad\quad 0 \le \theta \le 1 \tag{10.40}$$
$$t_\theta = (1-\theta)t_n + \theta t_{n+1} = t_n + \theta h.$$

They will be referred to as generalized mid-point schemes. The two families reduce to one if G is a linear function of the unknown (e.g. linear heat conduction): the so-called 'θ-method'. Particular cases are well known

$$\theta = \begin{cases} 0 & \text{Euler explicit (forward) scheme,} \\ \frac{1}{2} & \text{either trapezoidal or Crank–Nicolson (mid-point rule) scheme,} \\ \frac{2}{3} & \text{Galerkin scheme,} \\ 1 & \text{fully implicit or Euler backward scheme.} \end{cases}$$

All the schemes are consistent ones of the first order, except for $\theta = \frac{1}{2}$, in which case the two families are second-order accurate in the time step size.

Applied to system (10.37), the schemes (10.38) and (10.39) yield the following set of nonlinear algebraic equations, with $\mathbf{I}$ denoting the identity matrix

$$(\mathbf{I} + \theta h \mathbf{A}_{n+1}) T_{n+1} = [\mathbf{I} - (1-\theta)h\mathbf{A}_n] T_n + (1-\theta)h\mathbf{D}_n + \theta h \mathbf{D}_{n+1}, \quad (10.41)$$

for trapezoidal schemes

$$(\mathbf{I} + \theta h \mathbf{A}_\theta) T_{n+1} = [\mathbf{I} - (1-\theta)h\mathbf{A}_\theta] T_n + h\mathbf{D}_\theta \quad (10.42)$$

for mid-point schemes.

Thus every step involves the construction and solution of a new set of equations; therefore iterative solution techniques (except for $\theta = 0$) are required in which tangent or secant approximations may be used. The following linearization techniques have been proposed, however, in order to avoid an iterative solution technique:

1. A straight linearization over a time step by assuming

$$\mathbf{A}_{n+1} \cong \mathbf{A}_n, \qquad \mathbf{D}_{n+1} \cong \mathbf{D}(T_n, t_{n+1}) \quad (10.43)$$

for trapezoidal schemes
or

$$\mathbf{A}_\theta \cong \mathbf{A}_n, \qquad \mathbf{D}_\theta \cong \mathbf{D}(T_n, t_\theta) \quad (10.44)$$

for mid-point schemes. This corresponds obviously to a secant or initial load method of solution of the non-linear systems (10.41) or (10.42).

2. An extrapolation technique for the mid-point schemes family by taking

$$T_\theta = (1+\theta) T_n - \theta T_{n-1} \quad (10.45)$$

(obviously not valid for the first time step).

3. A predictor–corrector technique for the mid-point schemes family in which a predicted value T_{n+1}^* stems from (10.42), by assuming (10.44); a corrected value is then obtained by again solving (10.42) with

$$T_\theta^* = (1-\theta) T_n + \theta T_{n+1}^*.$$

The procedure thus requires the solution of two linear systems of equations per time step.

In two-step methods we consider the expression

$$(\theta + \tfrac{1}{2}) T_{n+2} - 2\theta T_{n+1} + (\theta + \tfrac{1}{2}) T_n + h\mathbf{A}_*[\beta_2 T_{n+2}$$
$$+ (1+\theta-2\beta_2) T_{n+1} + (\beta_2 - \theta) T_n] = h\mathbf{D}_* \quad (10.46)$$

Clearly every step implies the construction and solution of a new system of linear equations with the positive definite system matrix $[(\frac{1}{2} + \theta)\mathbf{I} + \beta_2 h\mathbf{A}_*]$. A different starting procedure is needed since the first application of (10.46) requires knowledge of T_1. One of the previous one-step schemes can be used for this purpose, but the problems associated with that changeover (e.g. step length change) will not be considered here.

Table 10.1 shows how most of the two-step schemes that have been presented in the literature are derived from (10.46) by particular choice of θ and β_2. It should be noted that some of these schemes are equivalent to particular one-step schemes: this is the case for the Dupont I scheme (Dupont et al. 1974) when $\alpha = \frac{1}{2}$ and the Dupont II scheme (10.50) when $\alpha = 0$, that both restore the Crank–Nicolson algorithm (scheme (10.42) with $\theta = \frac{1}{2}$). In addition a linearization of the unknown with $[t_n, t_{n+2}]$, i.e. the assumption that

$$T_{n+1} = (T_{n+2} + T_n)/2 \tag{10.47}$$

also restores particular one-step schemes, for instance:

1. The Crank–Nicolson scheme for (10.48) and (10.49) whatever the value of α
2. The mid-point scheme (10.42) with $\theta = \frac{3}{4}$ for (10.50) whatever the value of α
3. The fully implicit mid-point scheme ($\theta = 1$) for (10.51)

The two-step methods are widely used in solutions of problems involving phase change. The scheme used depends on the character of the phase change process. The most often used scheme in phase change problems is the Lee scheme.

Table 10.1 Linearized two-step schemes

θ	β_2	T_*	t_*	Algorithm name	Integration scheme	
0	$\frac{1}{3}$	T_{n+1}	t_{n+1}	Lee	$(\frac{1}{2}\mathbf{I} + \frac{1}{3}h\mathbf{A}_{n+1})T_{n+2}$ $= -\frac{1}{3}h\mathbf{A}_{n+1}T_{n+1} + (\frac{1}{2}\mathbf{I} - \frac{1}{3}h\mathbf{A}_{n+1})T_n + h\mathbf{D}_{n+1}$	(10.48)
0	α ($\alpha > \frac{1}{4}$)	T_{n+1}	t_{n+1}	Dupont I	$(\frac{1}{2}\mathbf{I} + \alpha h\mathbf{A}_{n+1})T_{n+2}$ $= -(1-2\alpha)h\mathbf{A}_{n+1}T_{n+1} + (\frac{1}{2}\mathbf{I} - \alpha h\mathbf{A}_{n+1})T_n + h\mathbf{D}_{n+1}$	(10.49)
$\frac{1}{2}$	$\frac{1}{2} + \alpha$ ($\alpha > 0$)	$\frac{3}{2}T_{n+1} - \frac{1}{2}T_n$	$t_{n+\frac{3}{2}}$	Dupont II	$[\mathbf{I} + (\frac{1}{2} + \alpha)h\mathbf{A}_*]T_{n+2}$ $= [\mathbf{I} - (\frac{1}{2} - 2\alpha)h\mathbf{A}_*]T_{n+1} - \alpha h\mathbf{A}_* T_n + h\mathbf{D}_*$	(10.50)
1	1	$2T_{n+1} - T_n$	t_{n+2}	Linearized fully implicit	$(\frac{3}{2}\mathbf{I} + h\mathbf{A}_*)T_{n+2} = 2T_{n+1}$ $+ \frac{1}{2}T_n + h\mathbf{D}_*$	(10.51)

10.1.4 Stability Analysis

Stability and accuracy properties of the schemes presented in Section 10.1.3 have been intensively described for both linear and nonlinear situations (see Kleiber and Służalec 1983a, for instance). Consider the case of the Euler explicit scheme

$$\dot{T}_{n+1} = \frac{1}{h}(T_{n+1} - T_n), \tag{10.52}$$

and the heat conduction matrix equation

$$\mathbf{K}T(t) + \mathbf{C}\,\dot{T}(t) = F(t). \tag{10.53}$$

To obtain the solution of the homogeneous equation we put

$$T(t) = e^{-\lambda t}\phi. \tag{10.54}$$

Inserting Eq. (10.54) into (10.53) we get

$$\mathbf{C}[-\lambda e^{-\lambda t}\phi] + \mathbf{K}[e^{-\lambda t}\phi] = 0, \tag{10.55}$$

and hence we obtain (since $e^{-\lambda t} \neq 0$)

$$\mathbf{K}\phi = \lambda \mathbf{C}\phi. \tag{10.56}$$

This equation is an n-order equation, where n signifies degrees of freedom of temperature.

This system of equations may be written in the following way

$$\mathbf{K}\phi = \mathbf{C}\phi\Lambda, \tag{10.57}$$

where ϕ is the C-orthonormal (i.e. $\phi^T\mathbf{C}\phi = \mathbf{I}$) eigenvector with components ϕ_1, $\phi_2, \ldots, \phi_n$ and Λ is a diagonal matrix containing the eigenvalues $\lambda_1, \lambda_2, \ldots, \lambda_n$.

Applying the transformation $T = \phi X$ (and $\dot{T} = \phi \dot{X}$) to Eq. (10.53) we can write it in spectral form as follows:

$$\mathbf{C}\phi\dot{X} + \mathbf{K}\phi X = F. \tag{10.58}$$

Inserting Eq. (10.57) into Eq. (10.58) we get

$$\mathbf{C}\phi\dot{X} + \mathbf{C}\phi\Lambda X = F, \tag{10.59}$$

which, multiplied by ϕ^T, gives

$$\phi^T\mathbf{C}\phi\dot{X} + \phi^T\mathbf{C}\phi\Lambda X = \phi^T F, \tag{10.60}$$

and finally, since ϕ is C-orthonormal

$$\dot{X} + \Lambda X = \phi^T F. \tag{10.61}$$

The above equations are independent and may be solved using an arbitrary numerical integration procedure.

Analysing the stability and accuracy of the iterative scheme, we see that integration of Eq. (10.61) is equal to the condition of integration of n independent differential equations of the first order in the base $\phi_1, \ldots, \phi_n$. Therefore, to analyse the stability, we can confine ourselves to a typical ith equation of the

form
$$\dot{x} + \lambda_i x = f, \qquad (10.62)$$
where
$$f = \phi_i^T F. \qquad (10.63)$$
Now we will go to the integration procedure using the step by step method. We know that
$$\dot{x}_{n+1} = \frac{1}{h}(x_{n+1} - x_n), \qquad (10.64)$$
hence it follows from Eq. (10.62) that
$$\frac{1}{h}(x_{n+1} - x_n) + \lambda_i x_{n+1} = f_{n+1}, \qquad (10.65)$$
or
$$x_{n+1} = Ax_n + Lf_{n+1}, \qquad (10.66)$$
where
$$A = \frac{1}{1 + \lambda_i h}, \qquad (10.67)$$
and
$$L = \frac{h}{1 + \lambda_i h}. \qquad (10.68)$$

A denotes the differential approximation operator and L the operator of heat transfer used in the ith independent nodal equation of the heat conduction.

From Eq. (10.66) we have the following condition for the time $t + m\Delta t$, m being an integer

$$x_{n+m} = A^m x_m + A^{m-1} Lf_{n+1} + \cdots + Lf_{n+m}. \qquad (10.69)$$

Consider the case when $L = 0$. We get

$$x_{n+m} = A^m x_n.$$

The integration method is unconditionally stable if A^m is bounded for large values of m (e.g. if $m \to \infty$) and arbitrary time step h. The stability condition can be described in the following way

$$A \leq 1. \qquad (10.70)$$

Hence by (10.67)

$$A = \frac{1}{1 + \lambda_i \Delta t} \leq 1, \qquad (10.71)$$

or

$$1 + \lambda_i \Delta t \geq 1. \tag{10.72}$$

Since all $\lambda_i > 0$, the integration algorithm is stable for $\Delta t > 0$ and the accuracy of the integration procedure depends on $\lambda_i \Delta t$.

10.2 Finite-Element Solution of Navier–Stokes Equations

Navier–Stokes equations (3.28)–(3.32) with boundary conditions given in Eqs. (3.33) and (3.34) form a complete set for the determination of the velocity, pressure and temperature fields in a moving fluid

$$\rho c \left(\frac{\partial T}{\partial t} + (v\boldsymbol{V})T \right) = \boldsymbol{V}(k\boldsymbol{V}T) + q_v, \tag{10.73}$$

$$\boldsymbol{V}v = 0, \tag{10.74}$$

$$\rho \left[\frac{\partial v}{\partial t} + v(\boldsymbol{V}v) \right] = \boldsymbol{V}\hat{\mathbf{p}} - \rho g \beta (T - T_0), \tag{10.75}$$

where

$$\hat{p}_{ij} = -p\delta_{ij} + \mu(v_{i,j} + v_{j,i}). \tag{10.76}$$

In matrix notation we can write

$$\boldsymbol{V}\hat{\mathbf{p}} = -\boldsymbol{V}p\mathbf{I} + \mu \boldsymbol{V} \Lambda^T v. \tag{10.77}$$

To apply the method of weighted residuals to this problem, we assume that within each element, a set of nodal points is established at which the dependent variables v_i, p and T are evaluated. It is assumed that the dependent variables can be expressed in terms of approximating functions by

$$v = \tilde{v}, \qquad p = \tilde{p}, \qquad T = \tilde{T},$$

where $\sim$ indicates an approximation. To derive a discrete set of equations appropriate for a particular element, the approximating functions are expressed by

$$\tilde{v} = \mathbf{N}^T v(t), \qquad \tilde{p} = \mathbf{M}^T P(t), \qquad \tilde{T} = \mathbf{H}^T T(t), \tag{10.78}$$

where $\mathbf{N}$ is a matrix, M, H are vectors of shape functions for the element and v, P, T are vectors of nodal point unknowns.

Applying the Galerkin method to Eqs. (10.73)–(10.77) we get the system equations

$$\left(\int_V H(\rho c) H^T \frac{\partial T}{\partial t} + (\rho c H \mathbf{N}^T v \boldsymbol{V} H^T T) \right) dV + \int_V \boldsymbol{V} H k \boldsymbol{V} H^T \, T \, dV$$

$$= \int_S Hqn \, dS + \int_V H q_v \, dV, \tag{10.79}$$

$$\left(\int_V M \boldsymbol{V} \mathbf{N}^T \, dV \right) v = 0. \tag{10.80}$$

$$\int_V \rho \left(\mathbf{N}\mathbf{N}^T \frac{\partial v}{\partial t} \right) dV + \left(\int_V \rho \mathbf{N}\mathbf{N}^T v \boldsymbol{\nabla} \mathbf{N}^T \, dV \right) v - \left(\int_V \boldsymbol{\nabla} \mathbf{N}^T \mathbf{M} \, dV \right) \mathbf{P}$$

$$+ \left(\int_V \mu \boldsymbol{\nabla} \boldsymbol{\Lambda}^T \, dV \right) v + \int_V \rho g \beta \mathbf{N} \mathbf{H}^T \, dV (T - T_0) + \int_S \mathbf{N} \tau \mathbf{n} \, dS = 0 \qquad (10.81)$$

From Eqs. (10.79)–(10.81) we get

$$\mathbf{C}\dot{T} + \mathbf{K}T + \mathbf{L}T = \mathbf{F}, \qquad (10.82)$$

$$\mathbf{A}v = 0, \qquad (10.83)$$

$$\mathbf{R}\dot{v} + \mathbf{E}v + \mathbf{D}v = \mathbf{G} + \mathbf{B}\mathbf{P}, \qquad (10.84)$$

where

$$\mathbf{C} = \int_V \rho c \mathbf{H}\mathbf{H}^T \, dV, \qquad (10.85)$$

$$\mathbf{K} = \int_V \boldsymbol{\nabla} \mathbf{H} k \boldsymbol{\nabla} \mathbf{H}^T \, dV, \qquad (10.86)$$

$$\mathbf{L} = \int_V \rho c \mathbf{H}\mathbf{N}^T v \boldsymbol{\nabla} \mathbf{H} \, dV, \qquad (10.87)$$

$$\mathbf{F} = \int_V \mathbf{H} q_v \, dV - \int_S \mathbf{H} q \mathbf{n} \, dS, \qquad (10.88)$$

$$\mathbf{A} = \int_V \mathbf{M} \boldsymbol{\nabla} \mathbf{N}^T \, dV, \qquad (10.89)$$

$$\mathbf{R} = \int_V \rho \mathbf{N}\mathbf{N}^T \, dV, \qquad (10.90)$$

$$\mathbf{E} = \int_V \rho \mathbf{N}\mathbf{N}^T v \boldsymbol{\nabla} \mathbf{N}^T \, dV, \qquad (10.91)$$

$$\mathbf{D} = \mu \int_V \boldsymbol{\nabla} \boldsymbol{\Lambda}^T \mathbf{N} \, dV, \qquad (10.92)$$

$$\mathbf{G} = - \int_V \rho g \beta \mathbf{N} \mathbf{H}^T \, dV (T - T_0) + \int_S \mathbf{N} \tau \mathbf{n} \, dS, \qquad (10.93)$$

$$\mathbf{B} = \int_V \boldsymbol{\nabla} \mathbf{N}^T \mathbf{M} \, dV. \qquad (10.94)$$

We recall briefly a solution procedure for the equations presented (Donea et al. 1981; Służalec 1988b, 1989a). Thermal field and fluid flow field can be found using the factorial step method. With the solution known everywhere at time $t_n = t_{n-1} + \Delta t$ we determine the solution of the fluid flow equation at the next time level t_{n+1}. In this method an intermediate velocity field v^* is calculated to satisfy only a discretized version of Eq. (10.84) without the pressure term. We have

$$\Delta v = \Delta t \, \mathbf{R}^{-1} (\mathbf{G} - \mathbf{E}v_n - \mathbf{D}v_n), \qquad (10.95)$$

$$v^* = v_n + \Delta v. \qquad (10.96)$$

The required velocity and pressure fields v_{n+1} and P_{n+1} are then obtained by adding to v^* the dynamic effect of pressure determined to ensure that the incompressibility condition remains satisfied

$$\Delta t \, \mathbf{E} \mathbf{R}^{-1} \mathbf{E} \mathbf{P}_{n+1} = -\mathbf{E} \mathbf{v}^*. \tag{10.97}$$

With the velocity and pressure at time t_{n+1} so determined the solution is completed by evaluating the temperature at this time from Eq. (10.82).

$$\Delta T = \Delta t \mathbf{C}^{-1} (\mathbf{F} - \mathbf{K} T - \mathbf{L} T). \tag{10.98}$$

$$T_{n+1} = T_n + \Delta T. \tag{10.99}$$

10.3 Modelling of the Phase Change Process

A lot of engineering problems involve a change of phase of materials from a liquid state to solid or vice versa. One obvious example is the welding of metals in which local melting and fusion of material is observed. One natural consequence of this process is the undesirable residual stresses induced in the structure. Another example is casting; the solidification of the molten metal often results in excessive distortion in the final shape of the structure.

Two major problems make the thermodynamical stress analysis of solids involving phase change difficult to handle. The first is the method of accounting for latent heat absorption or release during the phase change. The second one is the kinematics of the phase boundary (Meyer 1978; Crank 1981; O'Neil and Lynch 1981). The abrupt change of properties during phase change (Fig. 10.2) for most materials undoubtedly causes numerical instability in the analysis.

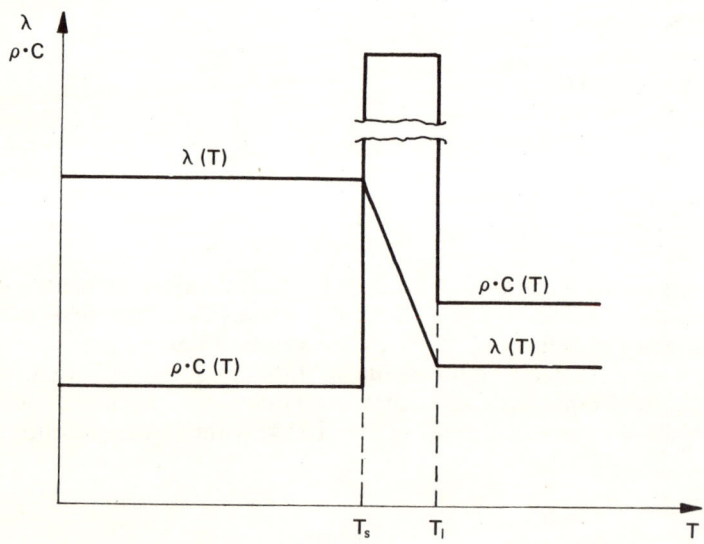

Figure 10.2. Thermomechanical properties in the range of ice–water phase change.

Attempts have been made to solve this type of problem by the classical method using the heat conduction equation with the latent heat and moving solid–liquid interface included in the boundary conditions. But only a limited number of problems involving simple geometries can be solved in this way. A different concept, recently adopted in finite-element formulations is the so-called enthalpy method.

In constructing the solution of problems involving phase change a possible approach is to track accurately the position of the phase boundary and then to solve Eq. (10.1) for the solid region and Eqs. (10.73)–(10.77) for the fluid region. Consider the case when the phase change process is modelled by a variant of the enthalpy method. In this method the phase change is assumed to occur over a temperature range and the associated latent heat effect is handled by increasing suitably the specific heat in this range (Comini et al. 1974; Morgan et al. 1978, 1980; Służalec, 1985, 1986a, 1987a,b, 1988a). Thus if the phase change is assumed to occur over the temperature interval $[T_s, T_l]$, where T_l is the liquidus and T_s the solidus temperature, then the specific heat c_A used in the calculation is defined by

$$c_A = c + \frac{[H(T - T_s) - H(T - T_l)]L}{\Delta T} \tag{10.100}$$

where H denotes the Heaviside function

$$H(t - a) = \begin{cases} 1 & (t > a), \\ 0 & (t \leq a), \end{cases} \tag{10.101}$$

L is the latent heat, and $\Delta T = T_l - T_s$ is the phase change interval.

When this method is applied to the analysis of pure materials, in which the phase change occurs at a specified temperature (i.e. $T_s = T_l$), a phase change time interval $\Delta t\, (\neq 0)$ must be assumed. It has been demonstrated (Comini et al. 1974; Gartling 1980; Morgan et al. 1980) that reasonable results can be obtained for problems involving conduction, provided that the right choice is made for the values of ΔT. However, for materials in which the phase change does occur over a reasonable temperature range this problem does not arise and the actual physical values of T_s and T_l can be used successfully. The enthalpy method can be used in conjuction with a fixed finite-element mesh. Gartling (1980) noted that such an approach was well suited to the computation of fluid motion except for the problem of the application of the non-slip (or zero fluid-velocity) boundary condition at the phase boundary. Exact application of this boundary condition at the phase change surface will not be possible as the fluid–solid interface will, in general, not coincide with the nodes of the finite-element mesh.

Gartling (1980) overcame this problem by using a smearing approach in which the viscosity was greatly increased in the phase change region. The problem can be also overcome (Morgan 1980) by adopting the simpler, but perhaps less accurate approach of reducing the nodal velocity to zero whenever the nodal temperature lies below the liquidus.

10.4 Examples of Thermal Problems

10.4.1 Heat Flow with Phase Change

We consider a heat exchange problem in the process of cryotherapy (Służalec and Muskalski 1985; Służalec 1986a, 1988b). Cryotherapy has a wide application in the treatment of many diseases. It is usually based on repeated touches of the cryoapplicator to the affected area of tissue; single, longer touches of the cryoapplicator are also used. The cryoapplicator should be maintained at a suitable temperature which depends on the type of treatment. In our example we analyse the change of temperature within the central part of the cornea with a diameter of 2 mm. The structural parameters of the cornea are illustrated in Fig. 10.3. As shown, the problem is axi-symmetrical. Thermal parameters of the cornea are listed in Table 10.2.

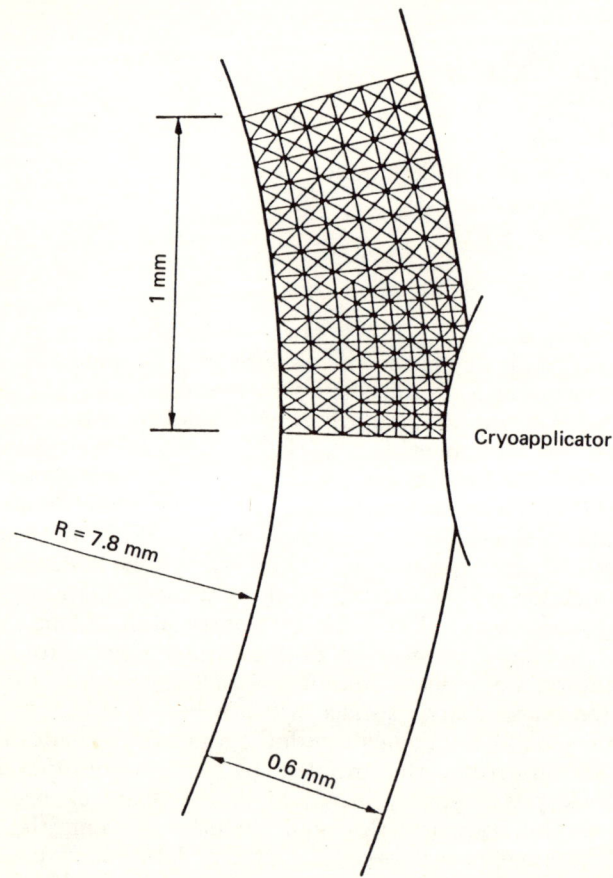

Figure 10.3. Finite-element model of the cornea.

Table 10.2 Thermal properties of cornea used for numerical analysis

State of region	Density (kg m^{-3})	Specific heat (J kg^{-1} K^{-1})	Thermal (W m K^{-1})	Latent heat (J kg^{-1})
Unfrozen	1000	4190	0.21	3.35×10^5
Frozen	900	2100	2.25	3.35×10^5

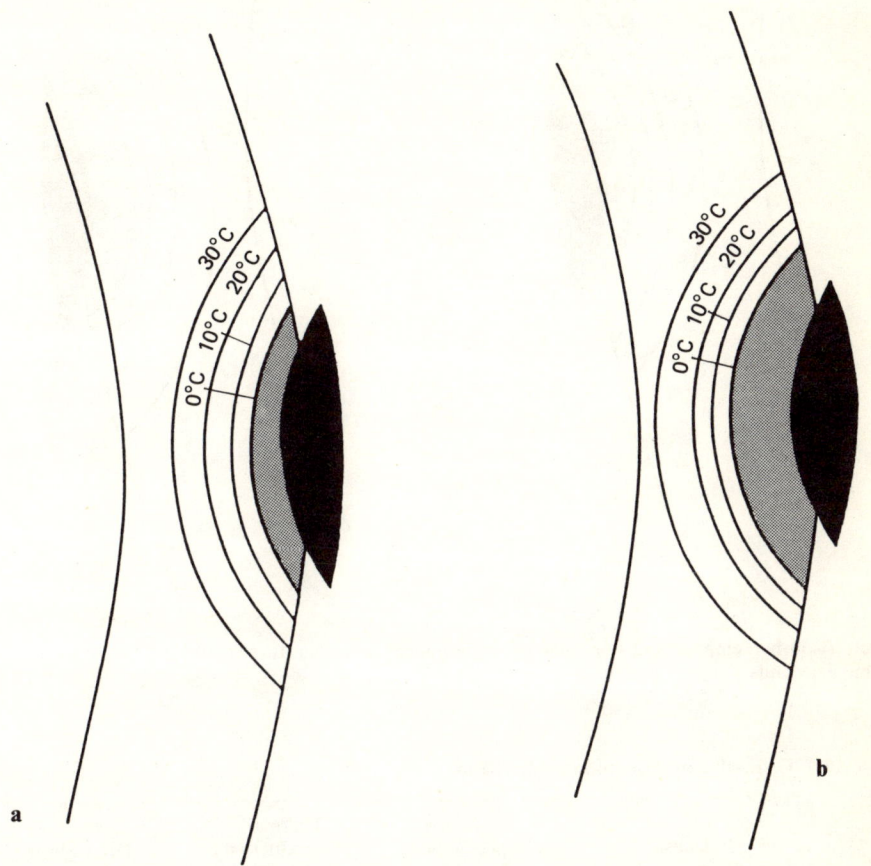

Figure 10.4a,b. Temperature distribution within cornea for cryoapplicator temperature $-70\,°C$: **a** time 4 seconds; **b** time 7 seconds.

Examples of the temperature distributions in the cornea four and seven seconds after the cryoapplications are illustrated in Figs. 10.4 and 10.5. The data shown refer to the dynamics of partial freezing of the cornea by a single touch of the cryoapplicator.

The next analysis is performed on the generation of frozen soil by means of artificial freezing in a groundwater flow field (Służalec 1989a). An array of

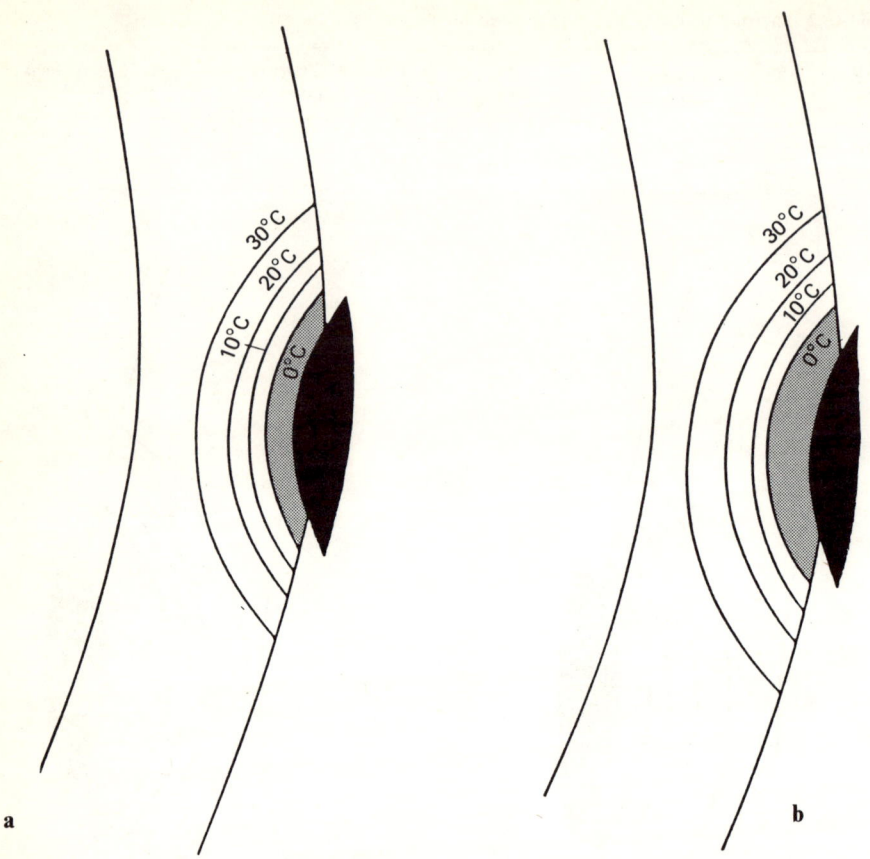

Figure 10.5a,b. Temperatures in cornea for cryoapplicator temperature $-40\,°C$: **a** time 4 seconds; **b** time 7 seconds.

Table 10.3 Thermal properties of saturated sand

Soil	Density (kg m^{-3})	Specific heat (J kg^{-1} K^{-1})	Thermal conductivity (W m K^{-1})	Latent heat (J kg^{-1})
Unfrozen	2020	1540	2.06	64 200
Frozen	1980	1120	2.69	64 200

freezing pipes is assumed to be normal to the flow direction. The thermal properties of the analysed soil are listed in Table 10.3. The finite-element mesh and boundary conditions are given in Fig. 10.6. The initial temperature of the ground is 20 °C, and the temperature of the freezing medium $-20\,°C$. A freezing

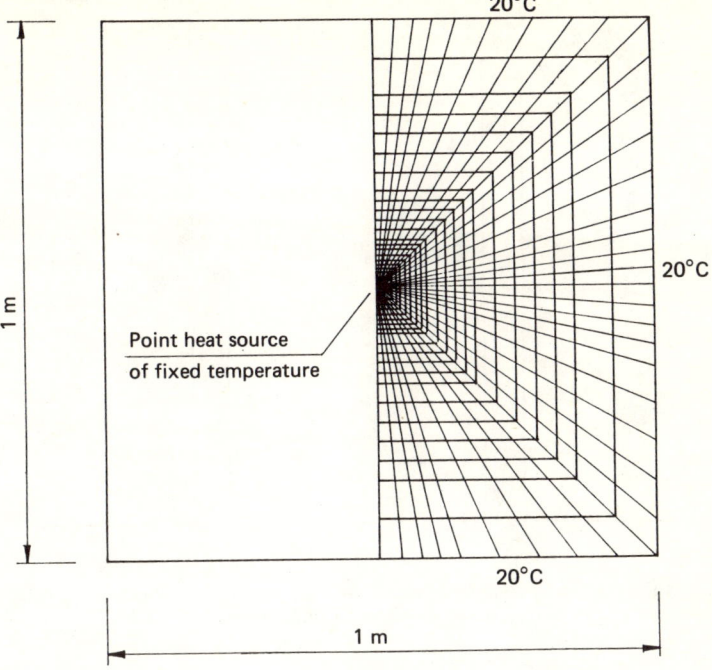

Figure 10.6. Finite-element mesh used for soil freezing.

pipe is considered as a point source. The results are shown in Figs. 10.7 through 10.9 for hydraulic gradients 0 (Fig. 10.7), 0.004 (Fig. 10.8) and 0.01 (Fig. 10.9) for the unfrozen regions. For the frozen domain we assumed that the velocity of the groundwater flow is zero.

10.4.2 Navier–Stokes Equations

At this point we present the application of numerical simulation to the analysis of electroslag welding (Służalec 1989b). A diagram of this process is presented in Fig. 10.10. A current is passed from a consumable electrode through a molten slag and molten metal pool. The 'Joule heating' in the slag causes the electrode to melt and the droplets thus formed pass through the slag and collect in the metal pool. The solidification of the pool causes the establishment of the joint connecting the two plates. Fluid motion in the system is represented by the fluid flow equations, which in essence express a balance between the rate of change of momentum within an infinitesimal fluid element and the sum of the net forces acting upon it. Fig. 10.11 presents the dimensions of the analysed region. The results of the analysis are presented in Figs. 10.12–10.14. The material parameters assumed for the analysis are: melting temperature of the electrode

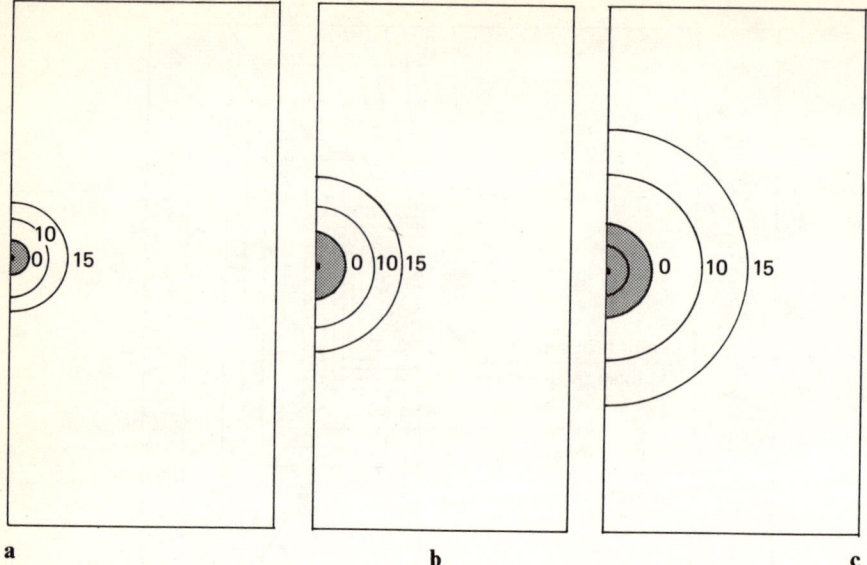

Figure 10.7. Progress of the freezing front and temperatures around the freezing pipe at hydraulic gradient 0, after: **a** 4 hours, **b** 16 hours, **c** 32 hours. (Temperatures in °C.)

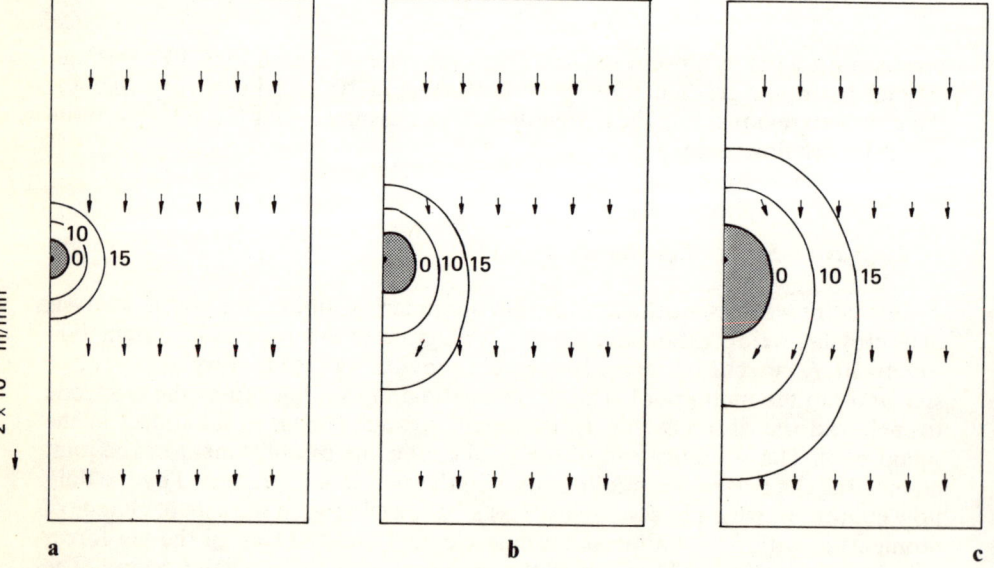

Figure 10.8a–c. Progress of the freezing front and temperatures around the freezing pipe at hydraulic gradient 0.004, after: **a** 4 hours; **b** 16 hours; **c** 32 hours. (Temperatures in °C.)

FINITE-ELEMENT SOLUTION 99

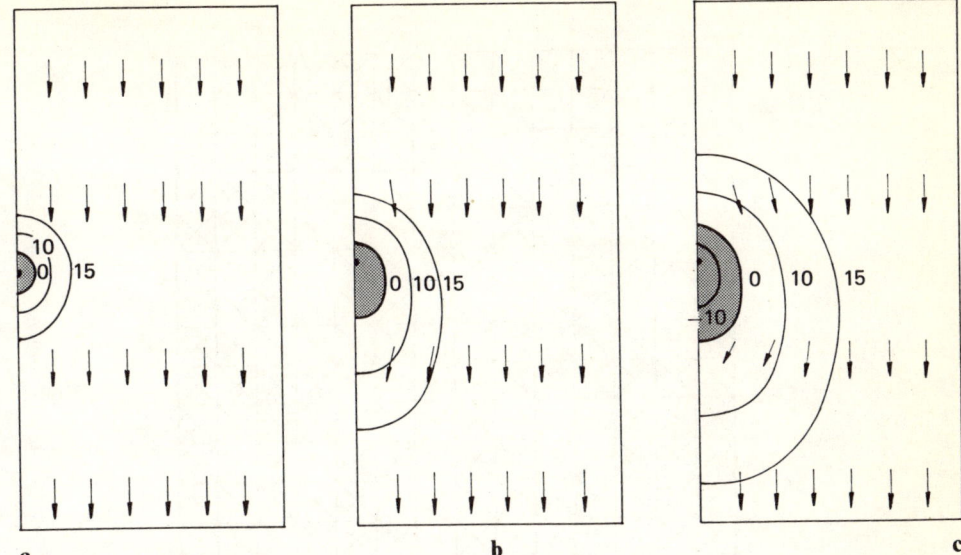

Figure 10.9a–c. Progress of the freezing front and temperatures around the freezing pipe at hydraulic gradient 0.01, after: **a** 4 hours; **b** 16 hours; **c** 32 hours. (Temperatures in °C.)

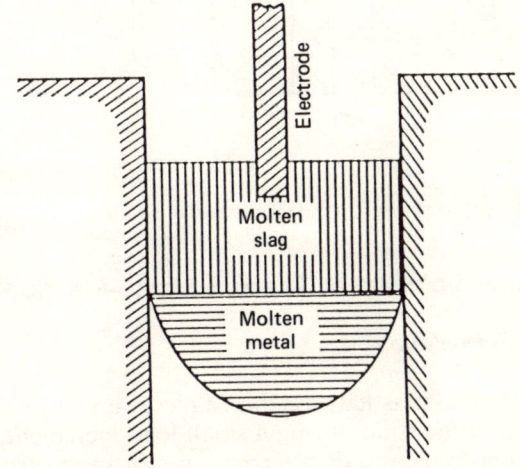

Figure 10.10. Scheme of electroslag welding.

1553 °C, melting temperature of the slag 1420 °C. The specific heat of the molten slag and metal, and other parameters are listed in Table 10.4. For a detailed description of boundary conditions as well as electro-mechanical forces in the regions we refer to the work of Służalec (1989b).

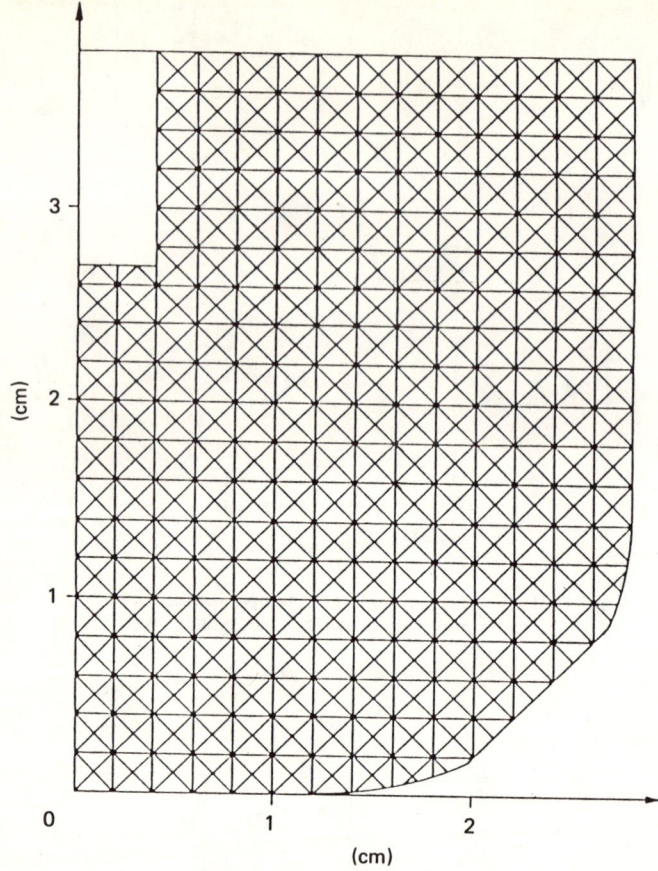

Figure 10.11. Finite-element mesh used to model electroslag welding.

10.5 Finite-Element Solution of Thermo-Elasto-Plastic Problems

10.5.1 Variational Formulation

The theory of incremental plasticity assumes a piecewise linear relation between the stress and strain in the solid during a small load increment. By a discretization procedure the incremental displacement vector in an element ΔU can be expressed in terms of the incremental nodal displacement vector Δu as follows

$$\Delta U(x) = \mathbf{N}(x)\Delta u, \tag{10.102}$$

where $\mathbf{N}(x)$ is the matrix of interpolation functions. The relation between the incremental displacement vector ΔU and the incremental strain vector $\Delta \varepsilon$ can be presented as

$$\Delta \varepsilon = \mathbf{B}(x)\Delta u. \tag{10.103}$$

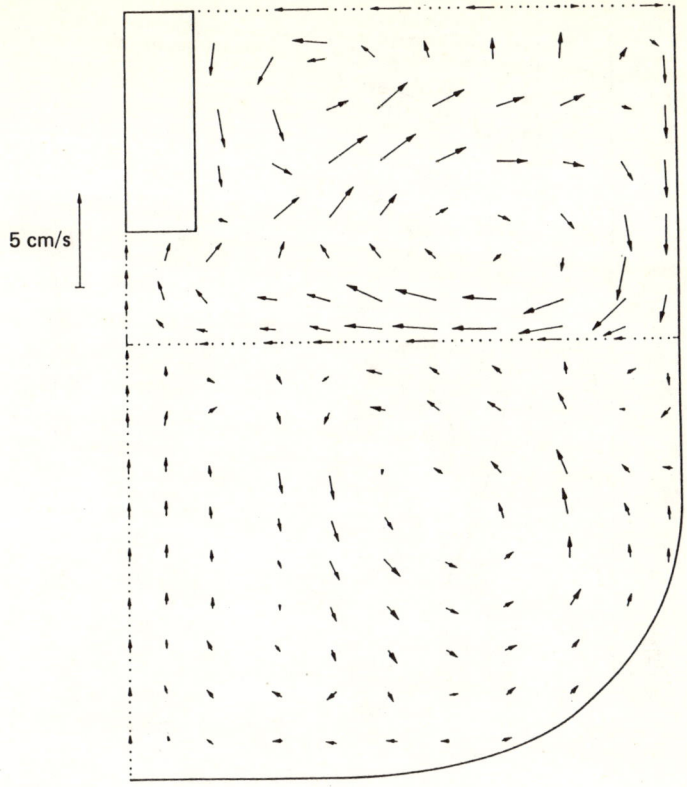

Figure 10.12. Velocity field for electroslag welding.

Consider the fundamental equation of thermo-elasto-plasticity in the case of isotropic hardening (9.15) written in matrix form

$$\Delta\boldsymbol{\sigma} = \mathbf{C}^e(\Delta\boldsymbol{\varepsilon} - \Delta\boldsymbol{\varepsilon}^T - \Delta\boldsymbol{\varepsilon}^{\bar{\varepsilon}}) = \mathbf{C}^e\,\Delta\boldsymbol{\varepsilon}', \qquad (10.104)$$

where

$$\Delta\boldsymbol{\varepsilon}' = \Delta\boldsymbol{\varepsilon} - \Delta\boldsymbol{\varepsilon}^T - \Delta\boldsymbol{\varepsilon}^{\bar{\varepsilon}}, \qquad (10.105)$$

$$\Delta\boldsymbol{\varepsilon}^T = \boldsymbol{\alpha}\,\Delta T + \frac{\partial \mathbf{E}}{\partial T}\boldsymbol{\sigma}\,\Delta T + \frac{\mathbf{C}^{ep-1}\mathbf{C}^e\boldsymbol{\sigma}}{S}\frac{\partial F}{\partial T}\,\Delta T, \qquad (10.106)$$

$$\Delta\boldsymbol{\varepsilon}^{\bar{\varepsilon}} = \frac{\partial \mathbf{E}}{\partial \bar{\varepsilon}}\boldsymbol{\sigma}\,\Delta\bar{\varepsilon} + \frac{\mathbf{C}^{ep-1}\mathbf{C}^e\boldsymbol{\sigma}}{S}\frac{\partial F}{\partial \bar{\varepsilon}}\,\Delta\bar{\varepsilon}. \qquad (10.107)$$

The incremental strain energy in the element is given by

$$\Delta U = \tfrac{1}{2}\int_V \Delta\boldsymbol{\varepsilon}'^T\,\Delta\boldsymbol{\sigma}\,dV. \qquad (10.108)$$

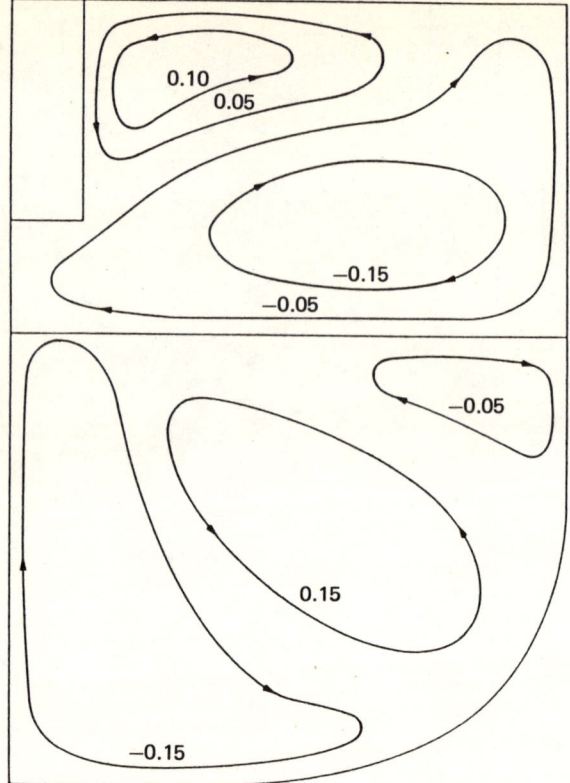

Figure 10.13. Streamlines (in kg s^{-1}) for the process of electroslag welding.

Table 10.4 Thermal parameters of molten metal and slag

	Specific heat (kcal kg^{-1} K^{-1})	Thermal conductivity (kcal m^{-1} s^{-1} K^{-1})	Density (g cm^{-3})	Viscosity (kg m^{-1} s^{-1})
Molten slag	0.2	0.0025	2.8	0.01
Molten metal	0.18	0.05	7.3	0.006

Substituting $\Delta\varepsilon$ given by (10.105)–(10.107) and $\Delta\sigma$ given by (10.104) in Eq. (10.108) we get the expression for ΔU in terms of Δu

$$\Delta U = \tfrac{1}{2} \Delta u^T \left(\int_V \mathbf{B}^T \mathbf{C}^{ep} \mathbf{B} \, dV \right) \Delta u - \Delta u^T \int_V \mathbf{B}^T \mathbf{C}^{ep} (\Delta \varepsilon^T + \Delta \dot{\varepsilon}) \, dV$$
$$+ \tfrac{1}{2} \int_V (\Delta \varepsilon^T + \Delta \dot{\varepsilon})^T \mathbf{C}^{ep} (\Delta \varepsilon^T + \Delta \dot{\varepsilon}) \, dV. \tag{10.109}$$

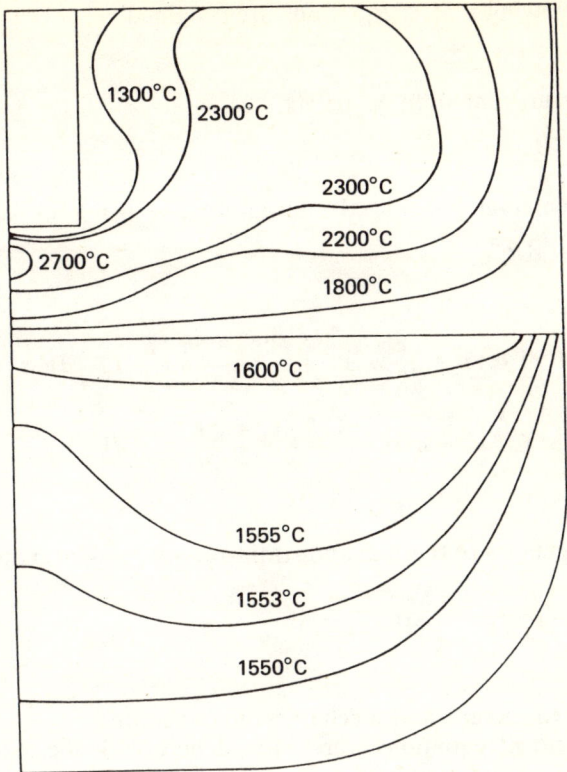

Figure 10.14. Computed temperatures for the process of electroslag welding.

Consider the following functional

$$\Delta \Pi = \Delta U - \int_V \Delta U^T \Delta f \, dV - \int_S \Delta U^T \Delta t \, dS \tag{10.110}$$

where Δf denotes incremental body forces and Δt incremental surface tractions. Finally we get

$$\Delta \Pi(\Delta u) = \tfrac{1}{2} \Delta u^T \left(\int_V \mathbf{B}^T \mathbf{C}^{ep} \mathbf{B} \, dV \right) \Delta u - \Delta u^T \int_V \mathbf{B}^T \mathbf{C}^{ep} (\Delta \varepsilon^T + \Delta \dot{\varepsilon}^{\dot{e}}) \, dV$$

$$+ \tfrac{1}{2} \int_V (\Delta \varepsilon^T + \Delta \dot{\varepsilon}^{\dot{e}})^T \mathbf{C}^{ep} (\Delta \varepsilon^T + \Delta \dot{\varepsilon}^{\dot{e}}) \, dV - \Delta u^T \int_V \mathbf{N}^T \Delta f \, dV$$

$$- \Delta u^T \int_S \mathbf{N}^T \Delta t \, dS. \tag{10.111}$$

Applying the variational principle with respect to u we get

$$\frac{\partial(\Delta \Pi)}{\partial(\Delta u)} = \int_V \mathbf{B}^T \mathbf{C}^{ep} \mathbf{B} \, dV) \Delta u - \int_V \mathbf{B}^T \mathbf{C}^{ep} (\Delta \varepsilon^T + \Delta \dot{\varepsilon}^{\dot{e}}) \, dV$$

$$- \int_V \mathbf{N}^T \Delta f \, dV - \int_S \mathbf{N}^T \Delta t \, dS = 0. \tag{10.112}$$

Finally the following matrix equations are obtained

$$\mathbf{K}\,\Delta u = \Delta R,\tag{10.113}$$

where $\mathbf{K}$ is the tangent stiffness matrix

$$\mathbf{K} = \int_V \mathbf{B}^T \mathbf{C}^{ep} \mathbf{B}\,dV,\tag{10.114}$$

ΔR is the incremental generalized external load vector

$$\Delta R = \Delta R^T + \Delta R^M,\tag{10.115}$$

where

$$\Delta R^T = \int_V \mathbf{B}^T \mathbf{C}^{ep}\left(\alpha\Delta T + \frac{\partial \mathbf{E}}{\partial T}\sigma\,\Delta T + \frac{\mathbf{C}^{ep^{-1}}\mathbf{C}^e\sigma}{S}\frac{\partial F}{\partial T}\Delta T\right)dV$$

$$+ \int_V \mathbf{B}^T \mathbf{C}^{ep}\left(\frac{\partial \mathbf{E}}{\partial \dot{\varepsilon}}\sigma\,\Delta\dot{\varepsilon} + \frac{\mathbf{C}^{ep^{-1}}\mathbf{C}^e\sigma}{S}\frac{\partial F}{\partial \dot{\varepsilon}}\Delta\dot{\varepsilon}\right)dV\tag{10.116}$$

includes all the effects of thermal and strain-rate-dependent material properties and

$$\Delta R^M = \int_V \mathbf{N}^T \Delta f\,dV + \int_S \mathbf{N}^T \Delta t\,dS\tag{10.117}$$

is the generalized incremental mechanical load vector.

The finite-element equations were derived here with the assumption of an isotropic hardening material model. One may find that similar equations can be readily made available for the kinematic hardening case, as the constitutive equations for both cases are virtually identical, except for the introduction of the translated stress defined by Eqs. (9.30) and (9.31).

10.5.2 Integration

The solution of thermo-elasto-plastic problems using the finite-element method is a complicated process. Certain numerical difficulties have been apparent for some classes of problem, particularly when gross section yielding has been involved. The method of dealing with the constitutive equations governing the material behaviour is very important. The original formulations concentrated on forward Euler schemes, but more recently the advantages of backward Euler schemes and the radial return algorithms have been realised. For incremental finite-element analysis, the previous equations of state have to be processed within each time step t_n during load incrementation and iteration. Given the values of stress, strain, displacement and temperature at t_n, a vector of nodal loads is used to find displacements and total strains at t_{n+1}. From these, using the constitutive law which expresses the plasticity rate equations, the stress and plastic strains can be evaluated to furnish a complete solution at t_{n+1}. Load residuals may be calculated to assess the current state of convergence and then to decide whether further iterations are required before completing the current load increment.

From the algorithmic point of view, the way $\dot{\varepsilon}^p$ is calculated is extremely important. Using the θ-method we have

$$\Delta \varepsilon^p = \Delta t \dot{\varepsilon}^p_{n+\theta} = \Delta t \lambda_{n+\theta} \mathbf{D} \sigma_{n+\theta}, \tag{10.118}$$

where

$$\sigma_{n+\theta} = (1 - \theta)\sigma_n + \theta \sigma_{n+1}, \tag{10.119}$$

$$\lambda_{n+\theta} = \lambda_{n+\theta}(\sigma_{n+\theta}, \dot{\varepsilon}_{n+\theta}, \dot{T}_{n+\theta}, \ldots), \tag{10.120}$$

D is the deviatoric stress matrix operator, and

$$X_{n+\theta} = X(\cdot, t + \theta \Delta t), \quad X_n = X(\cdot, t), \quad X_{n+1} = X(\cdot, t + \Delta t) \tag{10.121}$$

for arbitrary X (see Section 10.1.3).

The values of θ decide which type of procedure is relevant. Three main categories exist, differing by the direction of the flow vector at current time values thereby affecting accuracy and stability.
They are:

1. Tangent stiffness, radial corrector, $\theta = 0$
2. Mean normal, $\theta = \frac{1}{2}$
3. Elastic predictor, radial return, $\theta = 1$

The tangent stiffness–radial corrector ($\theta = 0$) method is the original method and has been used extensively. The stiffness matrix is frequently updated as $\mathbf{K}^T$ based on the pointwise elasto-plastic matrix $\mathbf{C}^{ep}$ which in turn is calculated by assuming the plastic flow in deviatoric stress space takes the direction of stress at t_n.

The mean normal ($\theta = \frac{1}{2}$) method uses a mean plastic flow direction, and conceptually lies between methods 1 and 3.

The elastic predictor–radial return ($\theta = 1$) algorithm assumes the stress can be corrected to lie on the yield surface by applying a fixed scaling factor to all components of deviatoric stress. The radial return method, first proposed by Wilkins (1964), has received much attention in recent years because of the high accuracy and quadratic convergence rate when used in conjunction with a consistent tangent stiffness matrix. The value of θ indicates the character of the method of numerical integration. Thus $\theta > 0$ implies an implicit method. For the HM criterion the method is unconditionally stable if $\theta < \frac{1}{2}$ but unconditionally stable otherwise.

10.5.3 Methods of Iterative Accumulation

An important factor in accumulating the state variables over iterations within a load increment is the point of reference of the data. Given an increment $\Delta t = t_{n+1} - t_n$, the plasticity algorithms produce in each time-step updates for displacements, total strains, stresses and plastic strains in that order. If each is a straight update of the values at t_n, then path-dependent updating takes place. If, however, all such updates are referred back to the start of the load increment, say t_N, with each value such as $d\sigma$ reflecting the rate value from that time, path-independent updating occurs.

Path dependence was invariably used with the older, radial correction type algorithms but because of the somewhat oscillatory and self-compensating behaviour of the increments in the main field values particularly over the early iterations of each load step, unfortunate effects could occur. Because plasticity is a discontinuous process, Gauss points which were nearly plastic had stresses oscillating into and out of a yielded state. The unnatural unloading associated with this has the effect of producing local point-wise divergence with tangent stiffness that could grow over several iterations to render a completely divergent solution.

Path independence is preferred with consistent tangent stiffness usage since such oscillatory effects are avoided. However, the field equations effectively involve larger increments and so the fundamental concept of plastic rate in these equations may not be entirely achieved. As in all numerical processes, sensibly-sized time (load) increments are implied.

10.5.4 Tangent Stiffness Matrices

The global tangent stiffness matrix $\mathbf{K}^T$ is derived from the usual elastic stiffness matrix plus updates of plastic information at each point of reference (Gauss point). This information appears in the point-wise modulus matrix $\mathbf{C}^{ep}$ of Eq. (9.14). These matrices are required to remain consistent with the numerical algorithm employed in integrating the plasticity rate equations so that the best (quadratic) convergence rate can be achieved. To do this the flow directions at point $n+1$ are used instead of those at point n. The latter indicates a continuum tangent stiffness matrix, with less than a quadratic convergence rate, whereas the former implies a consistent tangent stiffness matrix. A formal derivation of continuum and consistent operators is given by Simo and Taylor (1986).

The significance of these tangent operators varies with use of path-dependence or path-independence in strain updating. For the tangent stiffness–radial corrector integration scheme, the operator is continuum for path-independence but consistent for path-dependence. For the elastic predictor–radial return scheme, the operator is consistent for path-independence but continuum for path-dependence.

The consistent elastic-plastic matrix may be written as

$$\mathbf{C}^{ep} = \mathbf{C}^e - 2\gamma \mathbf{n}\mathbf{n}^T \tag{10.122}$$

where $\mathbf{C}^e$ is a 6×6 matrix c_{ij} with the only non-zero elements given by

$c_{ii} = K + \tfrac{4}{3}G\beta \quad (i = 1, 3), \qquad c_{ii} = G\beta \quad (i = 4, 6),$

$c_{ij} = K - \tfrac{2}{3}G\beta \quad (i, j = 1, 3 \text{ with } i \neq j), \qquad K$ is the bulk modulus,

$\mathbf{n} = \sqrt{\tfrac{3}{2}}\, s_{n+1}/\bar{\sigma}_{n+1}, \qquad \gamma = \dfrac{1}{1 + H'/3G} + \beta - 1$

where β is a parameter. For $\beta = 1$ the above matrix becomes the continuum one (Eq. (9.12)). The parameter β appearing here determines the distance of the point analysed from the yield surface. The elastic predictor–radial return algorithm can be used with the consistent elasto-plastic matrix and path-independent strain path updating.

For two-dimensional use, only the relevant rows and columns of $\mathbf{C}^{ep}$ are used. For plane stress, the third row and column is deleted using static condensation.

Consider the structural stiffness matrix $\mathbf{K}$ given by Eq. (10.114).

$$\mathbf{K}^\phi = \int_V \mathbf{B}^T \mathbf{C}^\phi \mathbf{B}\, dV. \tag{10.123}$$

The full elasto-plastic matrix $\mathbf{C}^{ep}$ gives a tangent stiffness matrix $\mathbf{K}^T$. This can be the elastic $\mathbf{C}^e$ matrix, when $\mathbf{K}^\phi$ becomes the initial stiffness matrix. This condition in turn is either consistent or continuum depending on the conditions alluded to earlier. In cases such as tangent stiffness–radial corrector methods, the use of $\mathbf{K}^T$ has been found commonly to cause divergence when sufficient plastic straining has accrued. This can be circumvented in practice with little loss in efficiency or accuracy, by using a partial tangent stiffness matrix $\mathbf{K}^\phi$ via equation (10.123) but with

$$\mathbf{C}^\phi = (1 - \phi)\mathbf{C}^e + \phi \mathbf{C}^{ep} \tag{10.124}$$

with ϕ chosen to lie in the interval $[0, 1]$. $\phi = 0$ gives $\mathbf{K}^0$ the initial stiffness matrix. In many cases when the use of $\mathbf{K}^T$ leads to divergence, a value of $\phi \cong 0.8$ may produce convergence.

10.6 Examples of Thermo-Elasto-Plastic Analyses

The first example of thermo-elasto-plastic analysis is the quenching problem of an aluminium cylinder. The numerical aspects of the strong material nonlinearities are of primary interest here. The quenching problem usually serves as an extreme test for residual stress analysis. The geometry of the cylinder is illustrated in Fig. 10.15 which is idealized by a slice of finite elements of unit height subjected to generalized plane strain conditions. The cylinder was heated to the temperature 540 °C and subsequently cooled on the outside surface by quenching with water $T_f = 22$ °C. The material properties are summarized in Table 10.5. The thermal computation gives the temperature history which is the

Table 10.5 Aluminium thermal and mechanical properties assumed for the analysis (units cm, W, g, J, °C, MPa)

$\lambda(T) = 2.13977 + 6.71948 \times 10^{-5} - 6.88114 \times 10^{-7}\, T^2$
$\rho(T) = 2.73022 - 2.15202 \times 10^{-4}\, T$
$c(T) = 0.89097 + 4.64526 \times 10^{-4}\, T$
$\alpha(T) = 22.32 \times 10^{-6} + 20.06 \times 10^{-9}\, T$
$H(T) = 16.553 - 1.6567 \times 10^{-2}\, T$
$H'(T) = 7079.7 - 7.0854 \times T$ (hardening modulus)
$E(T) = 70\,000 - 43.0395 \times (T - 20)$ (Young's modulus)
$v = 0.33$ (Poisson's ratio)

---

108 INTRODUCTION TO NONLINEAR THERMOMECHANICS

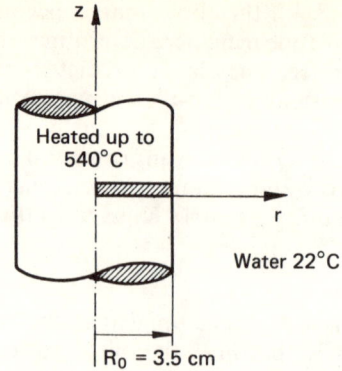

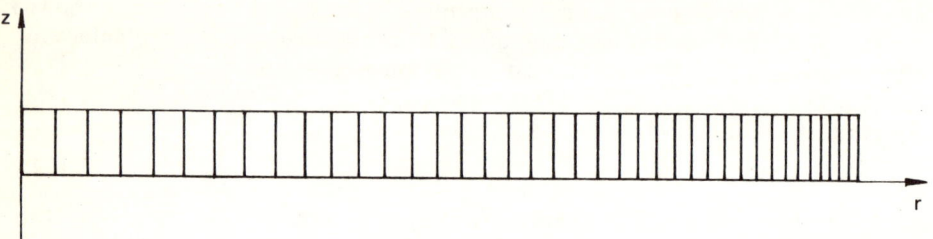

Figure 10.15. Finite-element model and boundary conditions for quenching problem.

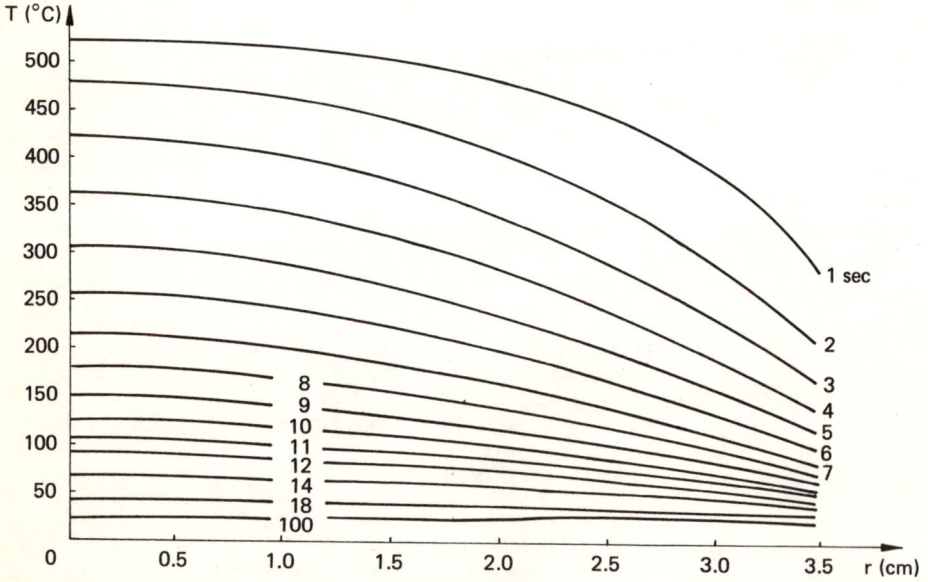

Figure 10.16. Temperature history for quenching problem.

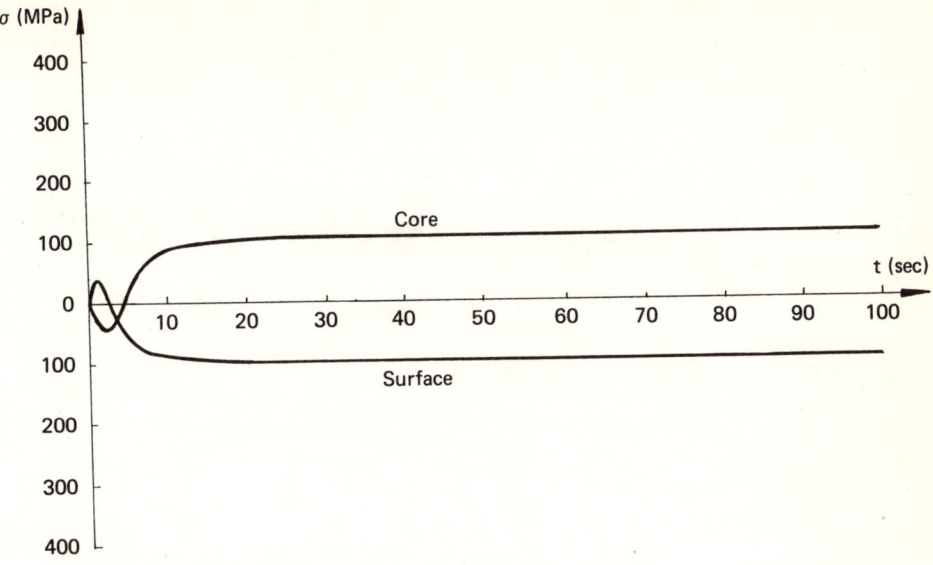

Figure 10.17. Stress σ_z at the core and on the surface.

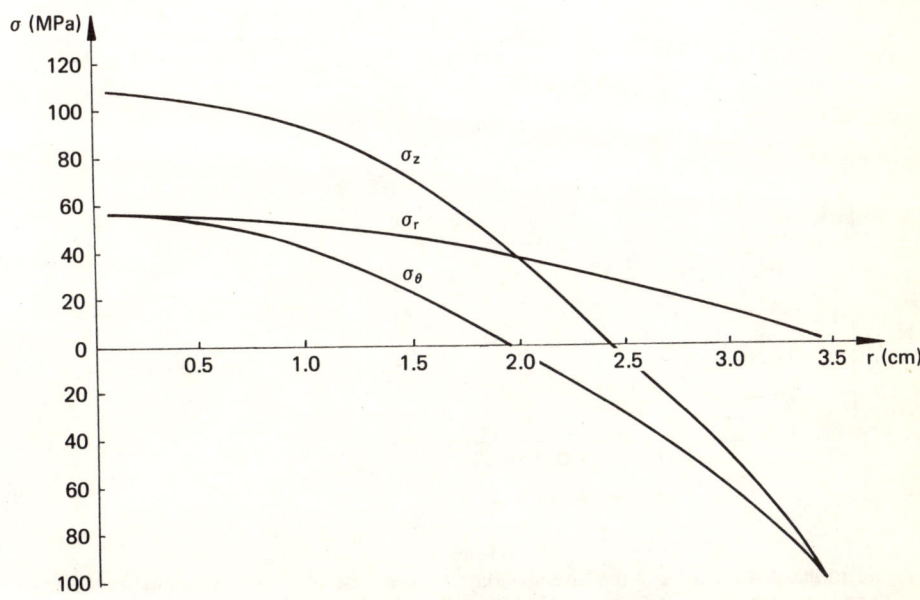

Figure 10.18. Residual stress distribution after 100 seconds of quenching.

110   INTRODUCTION TO NONLINEAR THERMOMECHANICS

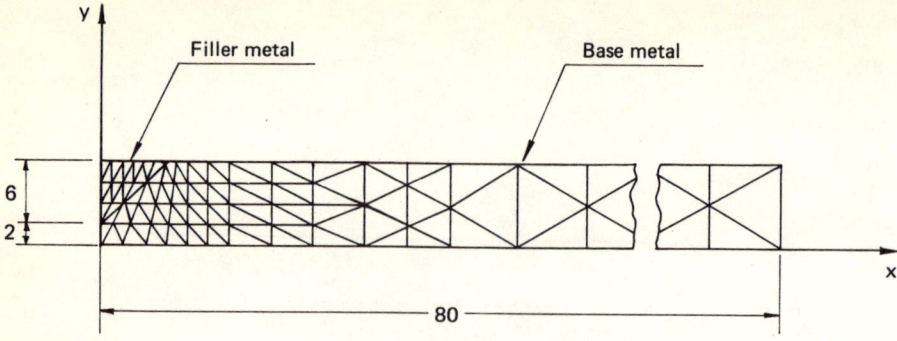

Figure 10.19. Finite-element mesh of the weld.

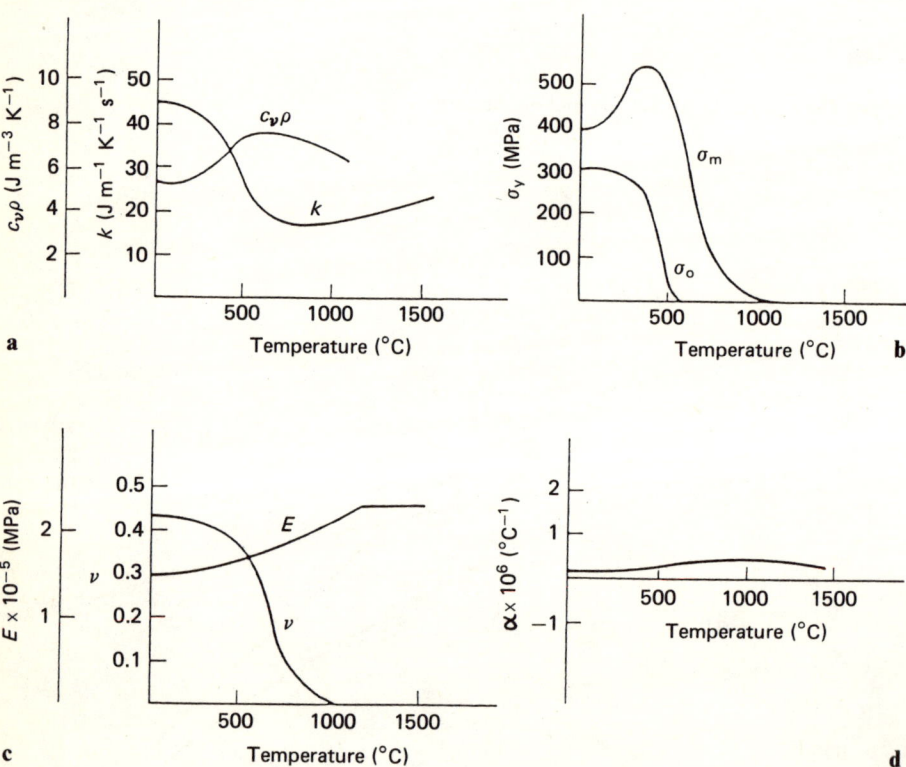

Figure 10.20a–d. Properties of the base material assumed for the analysis: **a** thermal conductivity $k(T)$ and heat capacity $c_v\rho(T)$; **b** initial $\sigma_0(T)$ and maximum yield limit $\sigma_m(T)$; **c** elastic modulus $E(T)$ and Poisson's ratio $\nu(T)$; **d** linear coefficient of thermal expansion $\alpha(T)$.

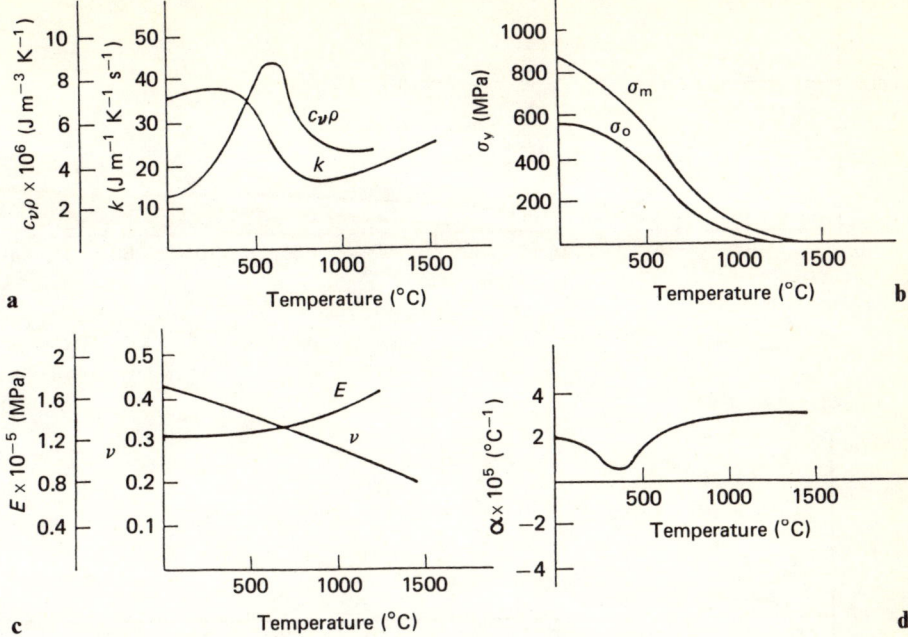

Figure 10.21a–d. Properties of the filler metal assumed for the analysis: **a** Thermal conductivity $k(T)$ and heat capacity $c_v\rho(T)$; **b** initial $\sigma_0(T)$ and maximum $\sigma_m(T)$ yield limit; **c** elastic modulus $E(T)$ and Poisson's ratio $\nu(T)$; **d** linear coefficient of thermal expansion $\alpha(T)$.

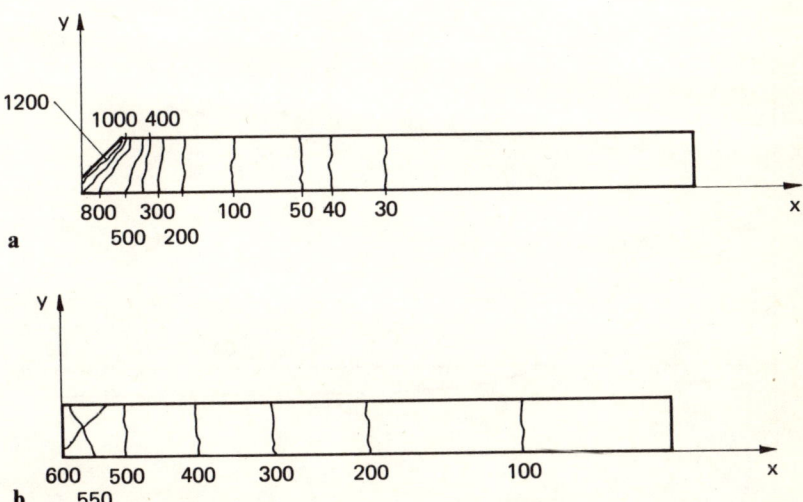

Figure 10.22a,b. Temperature distribution in the weld (°C): **a** heating phase $t = 5$ seconds; **b** cooling phase $t = 20$ seconds.

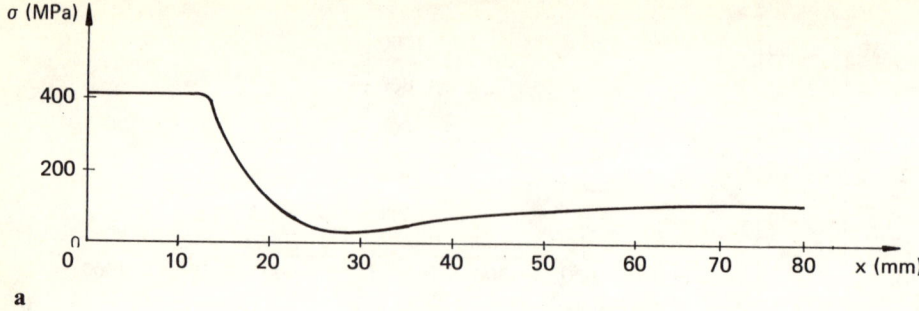

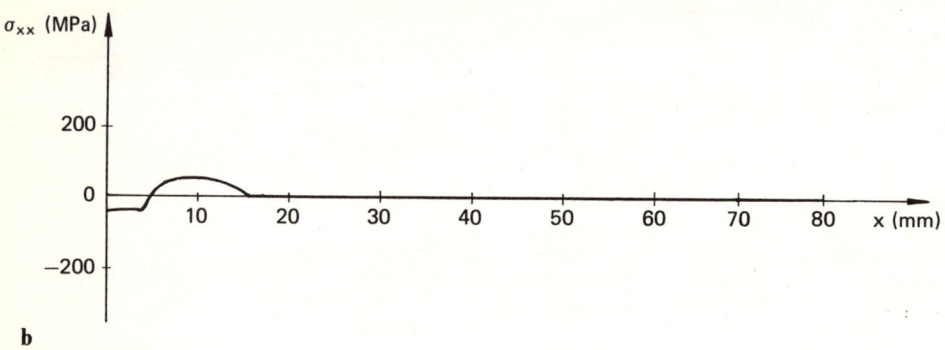

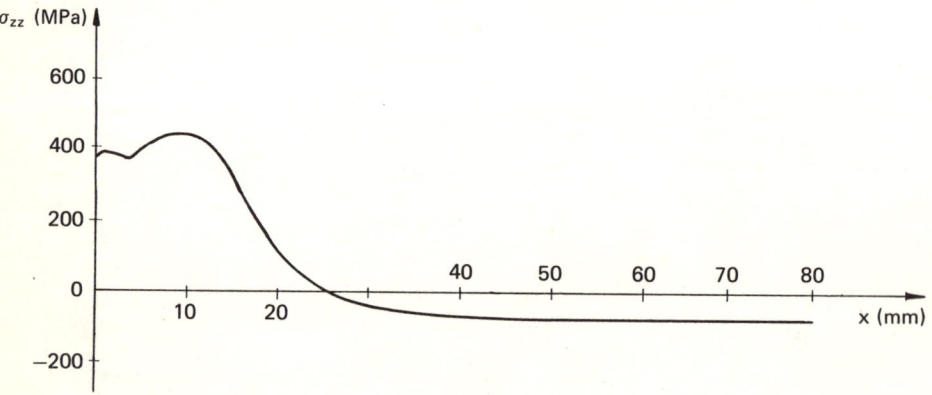

Figure 10.23a–c. Residual stress distributions at the top of the mid-section: **a** equivalent Huber–Mises stress; **b** transverse welding stress, **c** longitudinal welding stress.

only loading applied to the structure (Fig. 10.16). Fig. 10.17 shows σ_z stresses in the core and at the surface of the cylinder. Fig. 10.18 presents residual stress distribution after 100 seconds of quenching.

The second example is a thermo-elasto-visco-plastic welding stress analysis (cf. Służalec 1989c). A plane strain idealization quadrant at mid-section is presented in Fig. 10.19. The material properties assumed for computation are given in Figs. 10.20 and 10.21. The distribution of temperature contours in the mid-section for heating and cooling phases for two chosen times of the process is given in Fig. 10.22. Residual stress distributions at the top of the mid-section are presented in Fig. 10.23.

PART IV

CREEP

11 Theoretical Background to Creep

11.1 Creep and Relaxation Tests

Creep theory assumes a state equation which is more general than that which characterizes the theory of elasticity:

$$f(\varepsilon, \sigma, t, T) = 0. \tag{11.1}$$

At fixed temperature and prescribed stress history $\sigma(t)$ equation (11.1) determines strain variations in time $\varepsilon(t)$; analogously for prescribed strain history $\varepsilon(t)$, one can determine stress variations in time $\sigma(t)$. Among possible tests in this range one considers the creep test in constant stress condition Eq. (11.2) and a relaxation test in constant strain condition Eq. (11.3)

$$\sigma(t) = \sigma_1 H(t), \quad \varepsilon(t) = \sigma_1 C(t), \tag{11.2}$$

$$\varepsilon(t) = \varepsilon_1 H(t), \quad \sigma(t) = \varepsilon_1 E(t), \tag{11.3}$$

where $H(t)$ is the Heaviside function, and the functions $C(t)$ and $E(t)$ determining creep strain per unit applied stress and stress per unit applied strain, are called creep and relaxation functions respectively. Characteristic creep and relaxation curves are illustrated in Fig. 11.1. A typical creep curve is presented in Fig. 11.2. In this curve one may separate the ranges of primary creep (1), secondary creep (2) (with constant velocity), and tertiary creep preceeding rupture (3).

11.2 Creep at Constant Uniaxial Stress

Convenient approximations to the creep law (11.1) are obtained by separating stress, time and temperature influences

$$\varepsilon^c = f_1(\sigma) f_2(t) f_3(T). \tag{11.4}$$

11.2.1 Time Functions

For description of creep strain as a function of time in conditions of constant stress one uses the following:

$$\varepsilon^c = \beta t^{1/3} + kt \cong \beta t^{1/3} \quad -\text{Andrade (1910)}, \tag{11.5}$$

$\varepsilon^c = Ft^n$ — Bailey (1929), (11.6)

$\varepsilon^c = G(1 - e^{-qt}) + Ht$ — McVetty (1934), (11.7)

$\varepsilon^c = \sum a_i t^{n_i}$ — Graham and Walles (1955), (11.8)

$\varepsilon^c = \varepsilon_1 + A \lg t + Bt$ — Leaderman (1943), (11.9)

$\varepsilon^c = \varepsilon_1 + A \lg t$ — Philips (1950), (11.10)

$\varepsilon^c = \varepsilon_1 + \varepsilon t^n \quad (n < 1)$ — Findley (1944); Findley and Kholsa (1956); Findley and Peterson (1958). (11.11)

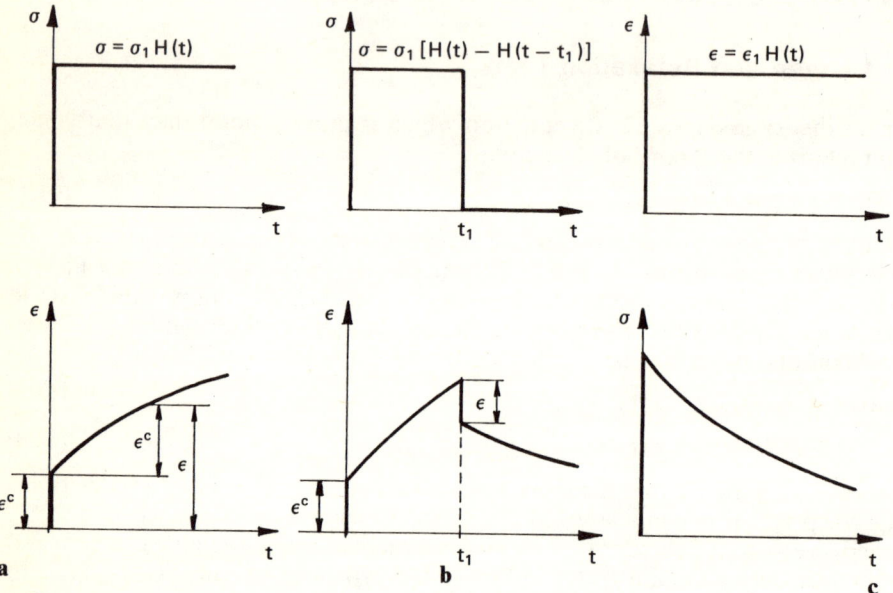

Figure 11.1a–c. Characteristic curves of creep and relaxation: **a** creep in constant stress condition; **b** creep during unloading; **c** relaxation in constant strain condition.

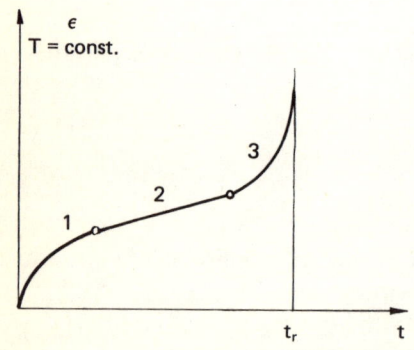

Figure 11.2. Primary (1), secondary (2) and tertiary (3) creep; t_r, rupture time.

11.2.2 Stress Functions

The best known relationships between creep rate and stress are given below:

$$\dot{\varepsilon}^c = K\sigma^m \quad (3 < m < 7) \quad \text{– Norton (1929), Bailey (1929),} \tag{11.12}$$

$$\dot{\varepsilon} = \frac{d}{dt}\left(\frac{\sigma}{\sigma_0^c}\right)^{n_0} + \left(\frac{\sigma}{\sigma_0^c}\right)^n \quad \text{– Odqvist (1966),} \tag{11.13}$$

$$\dot{\varepsilon} = D_1\sigma^{m_1} + D_2\sigma^{m_2} \quad \text{– Johnson et al (1963),} \tag{11.14}$$

$$\dot{\varepsilon}^c = B[\exp(\sigma/\sigma^c) - 1] \quad \text{– Soderberg (1936),} \tag{11.15}$$

$$\dot{\varepsilon}^c = A\sinh(\sigma/\sigma^c) \quad \text{– Nadai (1938), McVetty (1943),} \tag{11.16}$$

$$\dot{\varepsilon}^c = A[\sinh(\sigma/\sigma^c)]^m \quad \text{– Garofalo (1965),} \tag{11.17}$$

where $K, D_1, D_2, A, B, \sigma^c, \sigma_0^c, m, m_1, m_2, n, n_0$ are material constants.

11.2.3 Temperature Functions

The influence of temperature on creep is evident in two mechanisms: (a) temperature dependent material parameters, (b) structural changes in the material. An increase of creep rate is caused by temperature increase, in the course of increasing activation of structural elements. It opposes material hardening in the course of slip. In tests at temperatures $T < 0.4T_m$ (T_m = melting temperature) slip mechanisms are dominant during the deformation process (similar to plasticity). In the range $0.4T_m < T < 0.5T_m$ a significant growth in thermal mobility of dislocations occurs, causing obstacles to be bypassed by desertion of initial planes, thus decreasing the degree of material hardening. However, creep maintains its primary character (so-called cross-slip mechanism). In the temperature range $0.5T_m < T < 0.6T_m$ a new diffusive type mechanism can appear preventing further increase in the hardening effect. We refer to this as the secondary creep phenomenon (at constant creep rate). At temperatures $0.6T_m < T < 0.8T_m$ the diffusive character of creep leads to an increase of creep rate, characteristic of tertiary creep (Fig. 11.2).

For description of the influence of temperature on the creep rate the following equations have been proposed

$$\varepsilon^c = f[t\exp(-Q/RT)]f_1(\sigma) \quad \text{– Dorn and Tietz (1949), Dorn (1955),} \tag{11.18}$$

$$\varepsilon^c = [t\exp(-Q/RT)]^n f_1(\sigma) \quad \text{– Penny and Marriott (1971),} \tag{11.19}$$

$$\dot{\varepsilon}^c \simeq \exp\left(-\frac{Q - \gamma\sigma}{RT}\right) \quad \text{– Kauzman (1941),} \tag{11.20}$$

$$\dot{\varepsilon}^c \simeq \frac{\sigma}{T}\exp\left(-\frac{Q}{RT}\right) \quad \text{– Lifszic (1963),} \tag{11.21}$$

where Q is the activation energy, R is the gas constant and γ, n are material constants.

11.2.4 Stress and Time Functions

Common descriptions of stress and time influence on creep are the following

$$\left.\begin{array}{l}\varepsilon^c = g(\sigma)\,h(t),\\ \varepsilon^c = g_1(\sigma)\psi(t) + g_2(\sigma)t,\\ \varepsilon^c = g_1(\sigma)t^m + g_2(\sigma)t,\end{array}\right\} \quad \begin{array}{l}\text{- Malinin and Rżysko (1981);}\\ \text{Rabotnov (1966)}\end{array} \quad (11.22)$$

$$\varepsilon^c = \frac{C\sigma^2 t}{1 + k\sigma t} + g_2(\sigma)t \qquad \text{- Oding (1959),} \qquad (11.23)$$

$$\varepsilon = \frac{\sigma}{E} + k\sigma^p(1 - e^{-ct}) + L\sigma^q t \qquad \text{- Findley et al. (1976)} \qquad (11.24)$$

$$\varepsilon = \varepsilon_c \sinh(\sigma/\sigma_c) + \varepsilon^+ t^n \sinh(\sigma/\sigma^+) \quad \text{- Findley et al. (1948)} \qquad (11.25)$$

where C, k, E, L, ε_c, ε^+, σ_c, σ^+, n, p, q, c are functions which depend on temperature and $\psi(t)$ is a continuous, rapidly decreasing time function.

11.3 Creep Theories with Time-Dependent Uniaxial Stress

11.3.1 Total Strain Theory

The total strain theory assumes the existence, at any temperature, of a relationship between strain, stress and time (a surface for each T in the space ε, σ, t)

$$\phi_1(\varepsilon, \sigma, t, T) = 0. \qquad (11.26)$$

In Fig. 11.3 we present creep curves for stepwise, changing stress in total strain theory.

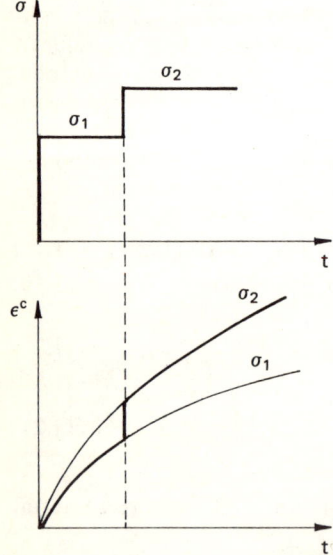

Figure 11.3. Creep curve for stepwise changing stress in total strain theory.

11.3.2 Time Hardening Theory

The time hardening theory, known also as the flow theory or second variant of the ageing theory, assumes the existence at a given temperature of a relationship between creep rate, stress and time (Davenport 1938; Kachanov 1960a)

$$\phi_2(\dot{\varepsilon}^c, \sigma, t, T) = 0. \tag{11.27}$$

Fig. 11.4 shows a graphical representation of the flow theory for stepwise changing stress.

11.3.3 Strain Hardening Theory

Strain hardening theory assumes the existence at a given temperature, of a relationship between creep rate, stress and creep strain (Ludwik 1909; Nadai 1938; Rabotnov 1966)

$$\phi_3(\dot{\varepsilon}^c, \sigma, \varepsilon^c, T) = 0, \tag{11.28}$$

$$\phi_3(\dot{p}, \sigma, p, T) = 0 \tag{11.29}$$

where $p = \varepsilon - \varepsilon^e$ means the nonelastic part of the deformation. Fig. 11.5 is an illustration of strain hardening theory.

A frequently used specialization of (11.29) which describes accurately the first segments of the creep curve is given by (Rabotnov 1966)

$$\dot{p}p^\alpha = f(\sigma) \tag{11.30}$$

where α is a constant.

The function $f(\sigma)$ is usually taken to be of the form

$$f(\sigma) = k\sigma^n \quad (n > 1 + \alpha), \tag{11.31}$$

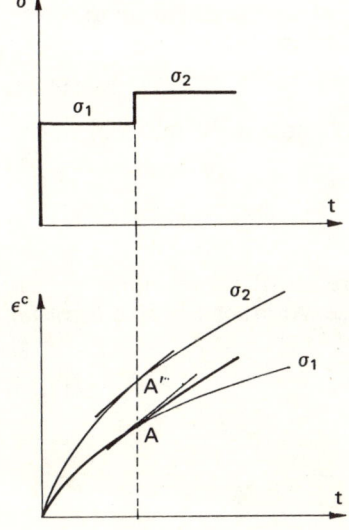

Figure 11.4. Creep curve for stepwise changing stress in time hardening theory.

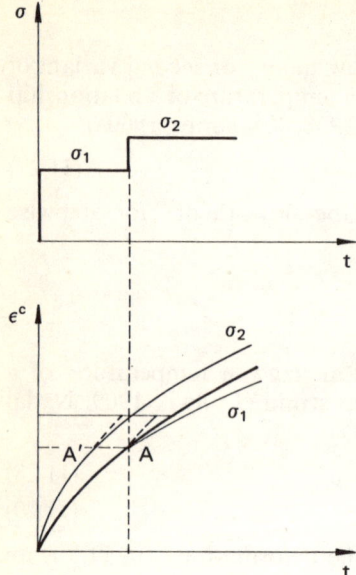

Figure 11.5. Creep curve for stepwise changing stress in strain hardening theory.

or

$$f(\sigma) = k\exp(\sigma/A), \tag{11.32}$$

or

$$f(\sigma) = k\left(2\sinh\frac{\sigma}{a}\right)^n. \tag{11.33}$$

Eq. (11.31) requires the limitation $n > 1 + \alpha$, because otherwise creep rate decreases when σ increases. For a proper description of both segments of the creep curve Rabotnov proposed a more general relationship than (11.30):

$$\dot{p}h(p) = f(\sigma) \tag{11.34}$$

where the function $h(p)$ characterizing strain hardening of the material can be assumed to be of the form (Szor 1958)

$$h(p) = \begin{cases} p^\alpha, & p \leq p_c, \\ p_c^\alpha, & p \geq p_c, \end{cases} \tag{11.35}$$

where p_c means such deformation, which is referred to the transition from the first, nonsteady, to the second, steady creep phase. Another possible form for $h(p)$ is

$$h(p) = (p^{-\alpha} + c)^{-1} \tag{11.36}$$

where α, c are constants which leads to

$$\frac{\dot{p}}{p^{-\alpha} + c} = f(\sigma). \tag{11.37}$$

For small deformations the function $h(p)$ determined by Eq. (11.36) changes with p^α and, as the deformation increases, it tends toward the constant value c, making possible a continuous description of the primary and secondary creep curve. Taking into account the temperature, Rabotnov proposed a relationship of the form

$$\dot{p}p^\alpha = f(\sigma)\exp(-Q/RT). \qquad (11.38)$$

11.3.4 Heredity Theory

In the group of heredity theories we include those that lead to a relationship between stress and strain in the form of integral equations of Volterra type (1951). The nonlinear Rabotnov heredity theory uses the integral equation in one of the forms given below

$$E(t)\phi[\varepsilon(t)] = \sigma(t) + \int_0^t K(t,\tau)\sigma(\tau)\,d\tau, \qquad (11.39)$$

or

$$E(t)\phi[\varepsilon(t)] = \sigma(t) + \int_0^t K(t-\tau)\sigma(\tau)\,d\tau. \qquad (11.40)$$

The kernel $K(t,\tau)$ or $K(t-\tau)$ determines the influence of loading impulse ($\sigma d\tau$) applied in time τ on strain magnitude at time of observation t. $\phi[\varepsilon(t)]$ is a nonlinear function of strain, which is to be determined on the basis of the σ–ε curve for instantaneous deformation (Fig. 11.6). In the case of unloading we put

$$\phi(t) = \phi'(\varepsilon') + E(\varepsilon - \varepsilon'). \qquad (11.41)$$

The kernel $K(t,\tau)$ is connected with the creep function $J(t,\tau)$ by the relationship

$$K(t,\tau) = -E(t)\frac{\partial J(t,\tau)}{\partial \tau}. \qquad (11.42)$$

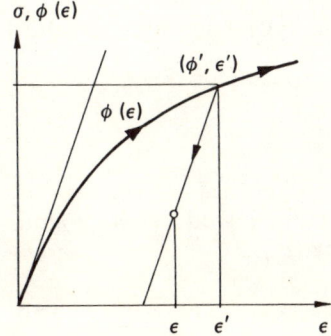

Figure 11.6. Determination of deformation function $\phi[\varepsilon(t)]$ in nonlinear heredity theory.

11.4 Creep Theories in Complex Stress State

11.4.1 Creep Theory of Deformational Type

Theories of deformational type are a straightforward transposition of the Nadai–Iljuszyn plasticity theory for description of creep strain:

$$e_{ij} = \frac{3}{2}\frac{\bar{\varepsilon}(\bar{\sigma},t)}{\bar{\sigma}}s_{ij}, \tag{11.43}$$

$$\varepsilon_{kk} = \frac{1}{3K}\sigma_{kk}. \tag{11.44}$$

Here the relation $\bar{\varepsilon} = \bar{\varepsilon}(\bar{\sigma},t)$ can be given by isochronous creep curves (Rabotnov 1948). In ageing theory Rabotnov (1948) proposed for $\bar{\varepsilon}(\bar{\sigma},t)$ equations similar to those for uniaxial states

$$\bar{\varepsilon} = \bar{\sigma}(t)\left[\frac{1}{3G} + \int_0^t \frac{S(\bar{\sigma})}{\bar{\sigma}}d\tau\right], \tag{11.45}$$

or

$$\bar{\varepsilon}\bar{\sigma}(t)\left[\frac{1}{3G} + \int_0^t B(\tau)[\bar{\sigma}(\tau)]^{m-1}d\tau\right]. \tag{11.46}$$

Substitution of (11.45) and (11.46) into (11.43) gives respectively

$$\varepsilon_{ij} = \frac{3}{2}\frac{s_{ij}}{E}\left[1 + E\int_0^t \frac{S(\bar{\sigma})}{\bar{\sigma}}d\tau\right], \tag{11.47}$$

$$\varepsilon_{ij} = \frac{3}{2}\frac{s_{ij}}{E}\left[1 + E\int_0^t B(\tau)[\bar{\sigma}(\tau)]^{m-1}d\tau\right]. \tag{11.48}$$

Rabotnov (1948) proposed a generalization of nonlinear ageing theory for multiaxial states:

$$\phi[\bar{\varepsilon}(t)] = \frac{\bar{\sigma}(t)}{3G} + \int_0^t K(t-\tau)\bar{\sigma}(\tau)d\tau. \tag{11.49}$$

It should be noted that theories of deformational type have little practical significance.

11.4.2 Flow Theories and Creep Potential

For generalization of Eq. (11.27) to multiaxial states one introduces, analogous to plasticity theory, the creep strain rate potential $\psi(\sigma_{ij})$

$$\dot{\varepsilon}_{ij}^c = \lambda\frac{\partial\psi}{\partial\sigma_{ij}} \tag{11.50}$$

or, eliminating the multiplier λ

$$\dot{\varepsilon}_{ij}^c = \frac{\partial \psi}{\partial \sigma_{ij}} \frac{\dot{\bar{\varepsilon}}^c}{\sqrt{\frac{2}{3} \frac{\partial \psi}{\partial \sigma_{mn}} \frac{\partial \psi}{\partial \sigma_{mn}}}}. \tag{11.51}$$

For isotropic material a good agreement with experiment is obtained by assuming ψ to be consistent with the HM criterion:

$$\psi(\sigma_{ij}) = s_{ij} s_{ij} - \tfrac{2}{3} \sigma_0^2. \tag{11.52}$$

Substituting (11.52) into (11.51) we get the flow theory equation

$$\dot{\varepsilon}_{ij}^c = \frac{3}{2} \frac{\dot{\bar{\varepsilon}}^c(\bar{\sigma})}{\bar{\sigma}} s_{ij}, \tag{11.53}$$

or more generally

$$\dot{\varepsilon}_{ij}^c = f(J_{2s}) s_{ij}. \tag{11.54}$$

Taking into account elastic strains and incompressibility we get

$$\dot{e}_{ij} = \frac{1}{2G} \dot{s}_{ij} + \frac{3}{2} \frac{\dot{\bar{\varepsilon}}^c(\bar{\sigma})}{\bar{\sigma}} s_{ij}. \tag{11.55}$$

For $\dot{\bar{\varepsilon}}^c(\bar{\sigma})$, Odqvist and Hult (1962) have proposed a generalization of laws of type (11.11), (11.12) for effective strain rate and effective stress as

$$\frac{\dot{\bar{\varepsilon}}^c}{\dot{\varepsilon}^c} = \left(\frac{\bar{\sigma}}{\sigma^c} \right)^n, \quad \dot{\bar{\varepsilon}}^c = \sqrt{\tfrac{2}{3} \dot{e}_{ij}^c \dot{e}_{ij}^c}, \quad \bar{\sigma} = \sqrt{\tfrac{3}{2} s_{ij} s_{ij}} \tag{11.56}$$

where $\dot{\varepsilon}^c$, σ^c, n are constants. Substituting the above relations into (11.53) and (11.55) we get finally (for $\dot{\varepsilon}^c = 1$)

$$\dot{\varepsilon}_{ij}^c = \dot{e}_{ij}^c = \frac{3}{2} \left(\frac{\bar{\sigma}}{\sigma^c} \right)^{n-1} \frac{s_{ij}}{\sigma^c} \tag{11.57}$$

and the Hooke–Norton equation

$$\dot{\varepsilon}_{ij} = \frac{1}{2G} \dot{s}_{ij} + \frac{3}{2} \left(\frac{\bar{\sigma}}{\sigma^c} \right)^{n-1} \frac{s_{ij}}{\sigma^c}. \tag{11.58}$$

In Eq. (11.58) one assumes that creep and elastic strains are associated with a deformation which is volume-preserving. In the absence of any constraints (11.58) is replaced by

$$\dot{\varepsilon}_{ij} = \frac{1+v}{E} \left(\dot{\sigma}_{ij} - \frac{v}{1+v} \dot{\sigma}_{kk} \delta_{ij} \right) + \frac{3}{2} \left(\frac{\bar{\sigma}}{\sigma^c} \right)^{n-1} \frac{s_{ij}}{\sigma^c}. \tag{11.59}$$

Further generalization of the Hooke–Norton equations made by Odqvist (1966) is based on replacement of the linear Hooke relationship by a nonlinear power law

$$\frac{\bar{\varepsilon}^e}{\varepsilon_0^c} = \left(\frac{\bar{\sigma}}{\sigma_0^c} \right)^{n_0}, \quad \bar{\varepsilon}^e = \sqrt{\tfrac{2}{3} e_{ij}^e e_{ij}^e}, \quad \bar{\sigma} = \sqrt{\tfrac{3}{2} s_{ij} s_{ij}} \tag{11.60}$$

and a similar law connecting the deviators

$$e_{ij}^e = g(J_{2s})s_{ij} \tag{11.61}$$

where $\varepsilon_0^c, \sigma_0^c, n_0$ are material constants.

By eliminating the function g, with assumption that elastic strains are associated with a deformation which is volume-preserving and setting $\varepsilon_0^c = 1, \dot{\varepsilon}_0^c = 1$ we get the Odqvist creep law

$$\frac{d\varepsilon_{ij}}{dt} = \frac{3}{2}\left\{\frac{d}{dt}\left[\left(\frac{\bar{\sigma}}{\sigma_0^c}\right)^{(n_0-1)}\frac{s_{ij}}{\sigma_0^c}\right] + \left(\frac{\bar{\sigma}}{\sigma^c}\right)^{(n-1)}\frac{s_{ij}}{\sigma^c}\right\}. \tag{11.62}$$

The above equations for $n_0 = 1$ and $\sigma_0^c = E$ are Hooke–Norton equations.

In equations of the flow theory one assumes the simplest possible relationship $\dot{\bar{\varepsilon}} = \dot{\bar{\varepsilon}}(\bar{\sigma})$. Full generalization of the flow theory for multiaxial states is obtained by assuming more general time-dependent strain hardening theory (Penny and Marriott 1971)

$$\dot{\bar{\varepsilon}} = f_1(\bar{\sigma})f_2(t)f_3(T). \tag{11.63}$$

Then

$$d\varepsilon_{ij}^c = f_1(\bar{\sigma})\frac{\partial \bar{\sigma}}{\partial \sigma_{ij}}\frac{df_2(t)}{dt}f_3(T)\,dt. \tag{11.64}$$

The equations of the flow theory so constructed take into account both the ageing effect and the temperature dependence.

The following equations of flow theory with ageing were used by Kachanov (1960).

$$\dot{\varepsilon}_{ij}^c = \frac{3}{2}\frac{f_1(\bar{\sigma})}{\bar{\sigma}}s_{ij}\dot{f}_2(t) \tag{11.65}$$

or, assuming $f_1(\bar{\sigma}) = \bar{\sigma}^m, \dot{f}_2(t) = B(t)$ and taking into account elastic strains

$$\dot{\varepsilon}_{ij} = \frac{1}{2G}\dot{s}_{ij} + \frac{3}{2}B(t)\bar{\sigma}^{m-1}s_{ij}. \tag{11.66}$$

Equations (11.66) are a direct generalization of the Hooke–Norton equations (11.59).

11.4.3 Generalization of Strain Hardening Theory

A generalization of the equations of the strain hardening theory (Section 11.3.3) is obtained by replacing the nonelastic part of the deformation p by the effective strain $\bar{p}$; the deformation rate $\dot{p}$ by the effective strain rate $\dot{\bar{p}}$; and the stress σ by the effective stress $\bar{\sigma}$. Instead of (11.30) we then get

$$\dot{\bar{p}}\bar{p}^\alpha = f(\bar{\sigma}) \tag{11.67}$$

where

$$\bar{p} = \sqrt{\tfrac{2}{3}\hat{p}_{ij}\hat{p}_{ij}}, \qquad \dot{\bar{p}} = \sqrt{\tfrac{2}{3}\dot{\hat{p}}_{ij}\dot{\hat{p}}_{ij}}, \qquad \bar{\sigma} = \sqrt{\tfrac{3}{2}s_{ij}s_{ij}} \tag{11.68}$$

and $\hat{p}_{ij}$, $\dot{\hat{p}}_{ij}$ are deviatoric creep strain and deviatoric creep strain rate components, respectively.

Assuming incompressibility and the similarity law connecting the deviatoric creep strain rate and deviatoric stress we get the expression

$$\dot{\hat{p}}_{ij} = \frac{3}{2} \frac{f(\bar{\sigma})}{\bar{p}^\alpha} \frac{s_{ij}}{\bar{\sigma}}. \tag{11.69}$$

In the case of using the state equation in more general form (11.34), we get

$$\dot{\bar{p}} h(\bar{q}) = f(\bar{\sigma}), \qquad \bar{q} = \int_0^t \sqrt{\tfrac{2}{3} \dot{\hat{p}}_{ij} \dot{\hat{p}}_{jj}} \, dt \tag{11.70}$$

which, by assuming the similarity of deviators and incompressibility leads to

$$\dot{\hat{p}}_{ij} = \frac{3}{2} \frac{f(\bar{\sigma})}{h(\bar{q})} \frac{s_{ij}}{\bar{\sigma}}. \tag{11.71}$$

In the above equation the length of the trajectory in creep strain space is used instead of effective creep strain $\bar{p}$.

12 Creep Rupture

12.1 Experimental Studies

Experimental studies involving creep rupture indicate the existence of two different rupture mechanisms. In ductile or transcrystalline rupture one observes a reduction of cross-section as a result of large creep strains caused by slip inside grains. Rupture takes place by propagation of cracks from the surface to the specimen interior along slip planes. Ductile rupture is characteristic of low temperatures and significant strain rates at high stress levels and large strains. Brittle or intercrystalline rupture is the result of material weakness caused by microcracks on grain boundaries because of slips between grains. Rupture takes place if the effective surface of the cross-section is reduced to the critical value. Brittle rupture is characteristic of high temperatures and small strain rates, for low stress level and small strains. The influence of temperature and strain rate on the type of rupture is illustrated in Fig. 12.1. The continuous line refers to large, and the dotted line to small strain rates. The characteristic creep rupture curves, showing the transition from ductile to brittle rupture, are presented in Fig. 12.1b (Grant and Bucklin 1965).

The most important problem of creep rupture theory is to extrapolate results of short-term creep tests (for instance 10^3 h) to the lifetime of the element (for instance 10^6 h). Larson and Miller (1952) initiated the development of methods

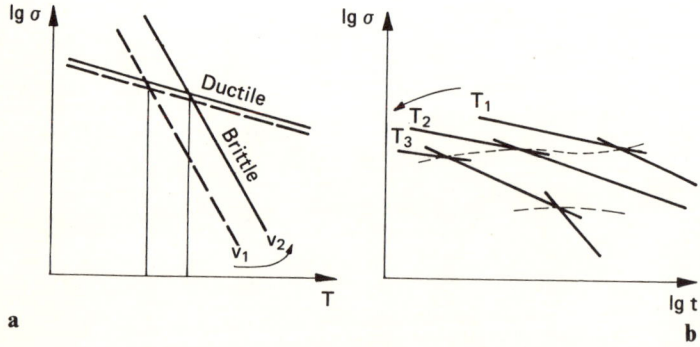

Figure 12.1a, b. Experimental results under creep rupture: **a** brittle and ductile ruptures; **b** rupture curves for various temperatures.

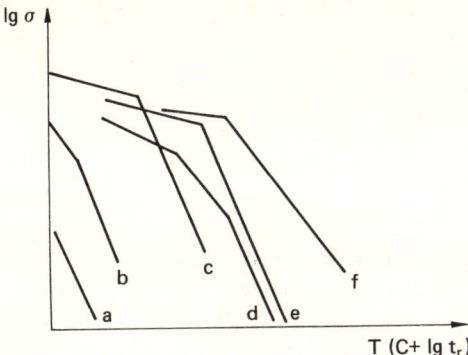

Figure 12.2a–g. Examples of universal creep curves as a function of the Larson–Miller parameter: **a** Al; **b** Al alloy; **c** Ti; **d** soft steel; **e** carbon steel; **f** stainless steel (after Penny and Marriott 1971).

with temperature and time parameters which make it possible to present a family of creep curves (with temperature as a parameter) in the form of one universal creep curve characteristic of a given material. It allows comparison of universal curves for various materials and valid extrapolations. Examples of universal creep curves for various materials as a function of the Larson–Miller parameter, $P = T(C + \lg t_r)$ are presented in Fig. 12.2. Other parameters used for creep studies and extrapolation techniques are given by Penny and Marriott (1971).

12.2 Ductile Rupture Theories

For determination of ductile rupture time t_r one has to integrate the chosen creep law in respect of the real stress σ and logarithmic strain ε^L. From definitions of σ and ε^L and from the condition of constant volume for uniaxial tension at constant applied nominal stress

$$\sigma = \frac{P}{A}, \quad \varepsilon^L = \ln \frac{L}{L_0}, \quad A_0 L_0 = AL, \quad \sigma_N = \frac{P}{A_0} = \sigma_1 \qquad (12.1)$$

one obtains the differential equation

$$\frac{1}{\sigma} \frac{d\sigma}{dt} = \frac{d\varepsilon^L}{dt} = f_1(\sigma) \qquad (12.2)$$

which has to be solved for a specified stress function $f_1(\sigma)$. The solution σ must satisfy the conditions

$$t = 0, \quad \sigma(0) = \sigma_1, \quad t = t_r, \quad \sigma(t_r) = \sigma_r. \qquad (12.3)$$

The rupture time is taken to be the time at which the real stress reaches the critical value σ_r. For most cases it is convenient to assume $\sigma_r \to \infty$. Hoff (1953) takes $f_1(\sigma)$ in the form given by Norton (1929) and Bailey (1929) in their creep law. With this assumption Eq. (12.2) becomes

$$\frac{1}{\sigma} \frac{d\sigma}{dt} = k\sigma^n. \qquad (12.4)$$

After integration and use of (12.3) one finds

$$t_r = \frac{1}{kn\sigma_1^n}\left[1 - \left(\frac{\sigma_1}{\sigma_r}\right)^n\right]. \tag{12.5}$$

Thus

$$t_r = t_H = \frac{1}{kn\sigma_1^n} = \lim_{\sigma_r \to \infty} \frac{1}{kn\sigma_1^n}\left[1 - \frac{\sigma_1}{\sigma_r}\right]. \tag{12.6}$$

Expressions (12.5) and (12.6) do not take into account the influence of the elastic strain on rupture time. The generalization of these results is obtained using the Hooke–Norton law (11.12) in application to large creep strains

$$\frac{1}{\sigma}\frac{d\sigma}{dt} = \frac{1}{E}\frac{d\sigma}{dt} + k\sigma^n. \tag{12.7}$$

After integration we have

$$t_r = \frac{1}{Ek(n-1)}\left(\frac{1}{\sigma_r^{n-1}} - \frac{1}{\sigma_1^{n-1}}\right) - \frac{1}{kn}\left(\frac{1}{\sigma_r^n} - \frac{1}{\sigma_1^n}\right). \tag{12.8}$$

Finally we get (Fig. 12.3)

$$t_r = \frac{1}{nk\sigma_1^n}\left[1 - \frac{n}{n-1}\frac{\sigma_1}{E} + \frac{n}{n-1}\left(\frac{\sigma_1}{E}\right)^n\right]. \tag{12.9}$$

A further generalization of the results (12.9) carried out by Odqvist (1961) is based on application of creep law (11.12) for large creep strains

$$\frac{1}{\sigma}\frac{d\sigma}{dt} = k_0\frac{d}{dt}(\sigma^{n_0}) + k\sigma^n. \tag{12.10}$$

After integration one obtains

$$t_r = \frac{1}{kn\sigma_1^n}\left(1 - \frac{nn_0 k_0}{n - n_0}\sigma_1^{n_0}\right) - \frac{1}{kn\sigma_r^n}\left(1 - \frac{nn_0 k_0}{n - n_0}\sigma_r^{n_0}\right). \tag{12.11}$$

Since if $t \to t_r$,

$$\sigma_r \to \sigma_r^* = (k_0 n_0)^{-1/n_0}, \quad \text{then} \quad d\sigma_r/dt \to \infty \tag{12.12}$$

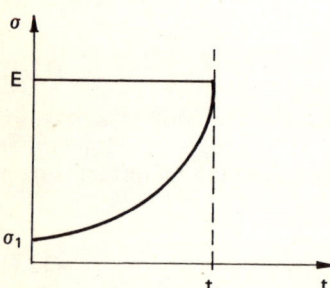

Figure 12.3. Ductile rupture on the basis of the Hooke–Norton law.

we get the following expression for rupture time

$$t_r = t_{01} = \frac{1}{kn\sigma_1^n}\left[1 - \frac{nn_0}{n-n_0}k_0\sigma_1^{n_0} + \frac{n_0}{n-n_0}\sigma_1(k_0n_0)^{\frac{n}{n_0}}\right]. \tag{12.13}$$

Assuming $\sigma_r = \sigma_{r_0} = \infty$ we obtain a slightly shorter rupture time

$$t_r = t_{02} = \frac{1}{kn\sigma_1^n}\left(1 - \frac{nn_0k_0}{n-n_0}\sigma_1^{n_0}\right), \quad t_{02} < t_{01}. \tag{12.14}$$

A more general concept is due to Rabotnov (1966). He formulated strain hardening theory (11.38) for large creep strains. Neglecting elastic strains and using logarithmic creep strains we get

$$p^L = \varepsilon^L - g(\sigma) = \ln\frac{\sigma}{\sigma_1} - g(\sigma), \tag{12.15}$$

where $g(\sigma)$ represents the plastic part of the strain. Substitution of the above into the equations of strain hardening theory (11.38) leads to

$$\frac{d\sigma}{dt}\left[\frac{1}{\sigma} - \frac{dg(\sigma)}{d\sigma}\right]h\left[\ln\frac{\sigma}{\sigma_1} - g(\sigma)\right] = f(\sigma). \tag{12.16}$$

After separation of variables and integration we obtain

$$t_r = t_R = \int_{\sigma_1}^{\sigma_r}\frac{d\sigma}{df(\sigma)}\left[1 - \frac{dg(\sigma)}{d\sigma}\right]h\left[\ln\frac{\sigma}{\sigma_1} - g(\sigma)\right]. \tag{12.17}$$

Using

$$h(p^L) = 1, \quad f(\sigma) = k\sigma^n, \quad g(\sigma) = k_0\sigma^{n_0}, \quad \sigma_r = \infty \tag{12.18}$$

we get after integration the Odqvist equation (12.14).

12.3 Brittle Rupture Theories

In order to describe the shrinking of the cross-section of a cylindrical body due to microcracks, Kachanov (1958, 1960a, 1961) proposed the continuity functions

$$\psi = \frac{A_{ef}}{A_0}, \quad \sigma_N = \frac{P}{A_0}, \quad \sigma = \frac{P}{A_{ef}} = \frac{\sigma_N}{\psi} \tag{12.19}$$

where A_0, A_{ef} are the initial and effective area of the cross-section respectively. Kachanov assumed that the rate of change of cross-section is a power function of real stress σ. The function ψ has to fulfil the following differential equation

$$\frac{d\psi}{dt} = -C\left(\frac{\sigma_N}{\psi}\right)^v \tag{12.20}$$

and conditions

$$\psi(0) = 1, \quad \psi(t_r) = 0 \tag{12.21}$$

where C and v are material constants (strongly temperature dependent), and t_r is the rupture time.

Under constant uniaxial nominal stress $\sigma_N = \sigma_1 = \text{const.}$ we get the expression for brittle rupture time (according to Kachanov)

$$t_r = t_K = \frac{1}{C(1+v)\sigma_1^v}. \tag{12.22}$$

For time-dependent nominal stress $\sigma_N = \sigma_N(t)$ equation (12.20) with conditions (12.21) gives

$$\int_0^1 \psi^v d\psi = -\int_0^{t_r} C[\sigma_N(t)]^v dt \tag{12.23}$$

or in equivalent form

$$\int_0^{t_r} \frac{dt}{t_K} = 1, \quad t_K = \frac{1}{C(1+v)[\sigma_N(t)]^v}. \tag{12.24}$$

In the case of uniaxial stress states Eq. (12.22) determines the initial rupture time which can be denoted by t_{r_1}. As a result of further stress redistribution the rupture is complete at time t_{r_2}, a little longer than the first one, $t_{r_2} > t_{r_1}$. In the case of multiaxial stress states Kachanov proposed the use of maximum principal stress $\sigma_{\max} = \sigma_I$ instead of σ_N in (12.20). Eq. (12.22) then becomes

$$t_I = \frac{1}{C(1+v)\sigma_I^v}. \tag{12.25}$$

Another, more universal concept is introduced by Sdobyriev (1959). It is based on replacing σ_I by σ_δ in (12.25), where

$$\sigma_\delta = \delta \sigma_I + (1-\delta)\bar{\sigma} \tag{12.26}$$

$\bar{\sigma}$ is the effective stress and δ is a material constant. According to Rabotnov (1968) for a wide class of materials δ is equal to $\frac{1}{2}$.

12.4 Rupture of Mixed Type

The theory of mixed brittle–ductile rupture takes into account the influence of shrinkage of cross-sections through geometric changes (large creep strains) and as a result of microcrack initiation. Kachanov (1958) proposed replacement of nominal stress $\sigma_N(t)$ in the creep law (12.20) by true stress $\sigma(t)$. Then the following two equations hold

$$\frac{1}{\sigma}\frac{d\sigma}{dt} = k\sigma^n, \tag{12.27}$$

$$\frac{d\psi}{dt} = -C\left(\frac{\sigma}{\psi}\right)^v. \tag{12.28}$$

By integrating (12.27) with the initial condition $\sigma(0) = \sigma_1$ one obtains

$$\sigma(t) = \sigma_1(1 - kn\sigma_1^n t)^{-1/n} = \sigma_1\left(1 - \frac{t}{t_H}\right)^{-1/n}. \tag{12.29}$$

Substituting the above expression into Eq. (12.28) and integrating using the conditions

$$\sigma(0) = \sigma_1, \quad \sigma(t_r) = \sigma_r, \tag{12.30}$$
$$\psi(0) = 1, \quad \psi(t_r) = 0, \tag{12.31}$$

gives

$$\int_0^1 \psi^v d\psi = -\int_0^{t_r} C\sigma_1^v \left(1 - \frac{t}{t_H}\right)^{-v/n} dt. \tag{12.32}$$

Finally the expression for brittle–ductile rupture time is arrived at

$$t_r = t_H\left[1 - \left(1 - \frac{n-v}{n}\frac{t_k}{t_H}\right)^{n/(n-v)}\right] \tag{12.33}$$

where

$$t_H = \frac{1}{kn\sigma_1^n}, \quad t_K = \frac{1}{C(1+v)\sigma_1^v}. \tag{12.34}$$

Eq. (12.34) is valid only for $t_r < t_H$, because otherwise ductile rupture appears. It leads to a limitation on the magnitude of initial nominal stress σ_1

$$\sigma_1 \leq \sigma_1^* = \left[\frac{C(1+v)}{k(n-\varphi)}\right]^{1/(n-v)}. \tag{12.35}$$

In the case $\sigma < \sigma_1^*$, the brittle–ductile rupture holds because of microcracks on grain boundaries and cross-section reduction. In the case $\sigma > \sigma_1^*$, the rupture mechanism is of ductile type and is caused only by cross-section reduction as a result of creep (Fig. 12.4).

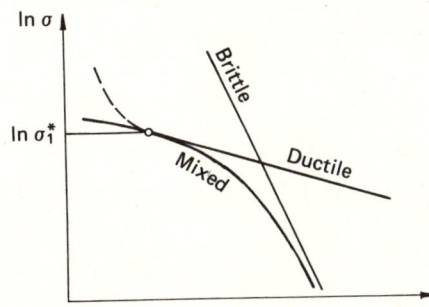

Figure 12.4. Mixed rupture.

13 Constitutive Equations for Thermo-Elasto-Plastic and Creep Analysis

Assume that for thermo-elasto-plastic and creep problems the following decompositions hold

$$\dot{\varepsilon}_{ij} = \dot{\varepsilon}^e_{ij} + \dot{\varepsilon}^T_{ij} + \dot{\varepsilon}^{e,T}_{ij} + \dot{\varepsilon}^{e,\bar{\varepsilon}}_{ij} + \dot{\varepsilon}^p_{ij} + \dot{\varepsilon}^c_{ij} \tag{13.1}$$

where $\dot{\varepsilon}^c_{ij}$ are components of the creep strain rate tensor.
If we denote

$$\dot{\varepsilon}^{tep}_{ij} = \dot{\varepsilon}^e_{ij} + \dot{\varepsilon}^T_{ij} + \dot{\varepsilon}^{e,T}_{ij} + \dot{\varepsilon}^{e,\bar{\varepsilon}}_{ij} + \dot{\varepsilon}^p_{ij} \tag{13.2}$$

then equation (13.1) may be expressed as

$$\dot{\varepsilon}_{ij} = \dot{\varepsilon}^{tep}_{ij} + \dot{\varepsilon}^c_{ij}. \tag{13.3}$$

The constitutive relation for thermo-elasto-plastic deformation with isotropic hardening is given in (9.15)

$$\dot{\sigma}_{ij} = C^{ep}_{ijkl}\dot{\varepsilon}^{tep}_{kl} - C^{ep}_{ijkl}(\alpha_{kl}\dot{T} + D^T_{kl}\dot{T} + D^{e,\bar{\varepsilon}}_{kl}(\bar{\varepsilon})^{\cdot})$$
$$- \frac{C^e_{ijkl} S_{kl}}{S}\left(\frac{\partial F}{\partial T}\dot{T} + \frac{\partial F}{\partial \bar{\varepsilon}}(\bar{\varepsilon})^{\cdot}\right). \tag{13.4}$$

Rearrangement of the above equation to express $\dot{\varepsilon}^{tep}_{ij}$ in terms of $\dot{\sigma}_{ij}$ gives

$$\dot{\varepsilon}^{tep}_{kl} = (C^{ep}_{ijkl})^{-1}\dot{\sigma}_{ij} + (\alpha_{kl}\dot{T} + D^T_{kl}\dot{T} + D^{e,\bar{\varepsilon}}_{kl}(\bar{\varepsilon})^{\cdot})$$
$$+ \frac{(C^{ep}_{ijkl})^{-1} C^e_{ijkl} S_{kl}}{S}\left(\frac{\partial F}{\partial T}\dot{T} + \frac{\partial F}{\partial \bar{\varepsilon}}(\bar{\varepsilon})^{\cdot}\right) \tag{13.5}$$

By (13.1), (13.2), (13.4) and (13.5) we get finally

$$\dot{\sigma}_{ij} = C^{ep}_{ijkl}\dot{\varepsilon}_{kl} - C^{ep}_{ijkl}(\alpha_{kl}\dot{T} + D^T_{kl}\dot{T} + D^{e,\bar{\varepsilon}}_{kl}(\bar{\varepsilon})^{\cdot} + \dot{\varepsilon}^c_{kl})$$
$$- \frac{C^e_{ijkl} S_{kl}}{S}\left(\frac{\partial F}{\partial T}\dot{T} + \frac{\partial F}{\partial \bar{\varepsilon}}(\bar{\varepsilon})^{\cdot}\right). \tag{13.6}$$

The phenomenon of stress relaxation in a structure is indicated in Eq. (13.6) by the negative sign of $\dot{\varepsilon}^c_{kl}$.

14 Finite-Element Formulation

14.1 Matrix Equation for Thermo-Elasto-Plastic and Creep Problems

By comparing the constitutive equation (9.15) for the thermo-elasto-plastic problem with those given in (13.6) where the creep effect has been included it appears that the only difference between these two equations is the additional term $\dot{\varepsilon}_{ij}^c$ in the second equation. Thus, by the procedure described in Section 10.5.1 one may readily conclude that the same element equation can be used for the present case with only one minor modification to the thermomechanical load matrix vector for the creep effect. The finite-element matrix equations can be shown to be

$$\mathbf{K}\Delta u = \Delta R \tag{14.1}$$

where

$$\mathbf{K} = \int_V \mathbf{B}^T \mathbf{C}^{ep} \mathbf{B}\, dV, \tag{14.2}$$

$$\Delta R = \Delta R^T + \Delta R^M + \Delta R^C, \tag{14.3}$$

$$\Delta R^T = \int_V \mathbf{B}^T \mathbf{C}^{ep} \left(a\Delta T + \frac{\partial \mathbf{E}}{\partial T}\sigma\Delta T + \frac{(\mathbf{C}^{ep})^{-1}\mathbf{C}^e}{S}\sigma\frac{\partial F}{\partial T}\Delta T \right) dV$$

$$+ \int_V \mathbf{B}^T \mathbf{C}^{ep} \left(\frac{\partial \mathbf{E}}{\partial \dot{\varepsilon}}\sigma\Delta\dot{\varepsilon} + \frac{(\mathbf{C}^{ep})^{-1}\mathbf{C}^e\sigma}{S}\frac{\partial F}{\partial \dot{\varepsilon}}\Delta\dot{\varepsilon} \right) dV, \tag{14.4}$$

$$\Delta R^M = \int_V \mathbf{N}^T \Delta f\, dV + \int_S \mathbf{N}^T \Delta t\, dS \tag{14.5}$$

and

$$\Delta R^C = \int_V \mathbf{B}^T \mathbf{C}^{ep} \Delta \varepsilon^c\, dV. \tag{14.6}$$

The vector ΔR^c is the additional load vector induced by the creep effect. Comparison of (10.113) and (14.1) indicates that the solution of thermo-elasto-plastic and creep problems differs from that of thermo-elasto-plastic analysis only by an additional creep load vector in the overall nodal force matrices.

14.2 Remarks on Solution Procedures

The most popular solution procedure for solving creep problems is based on a one-parameter integration method for ordinary differential equations known as the θ-method (see Sections 10.1.3 and 10.5.2). This method is recommended for simultaneous solution of creep and plasticity problems. By a method analogous to plastic strain analysis, the creep rate $\dot{\varepsilon}^c$ at time $t + \Delta t$ is first decomposed as

$$\dot{\varepsilon}^c_{n+1} = \dot{\varepsilon}^c_n + \Delta \varepsilon^c \tag{14.7}$$

with the assumption that, for an arbitrary time step n

$$\dot{\varepsilon}^c_n = \gamma_n \mathbf{D} \sigma_n, \tag{14.8}$$

where $\mathbf{D}$ is the deviatoric stress operator

$$\Delta \varepsilon^c = \Delta t \dot{\varepsilon}^c_{n+\theta} = \Delta t \gamma_{n+\theta} \mathbf{D} \sigma_{n+\theta}, \tag{14.9}$$

$$\sigma_{n+\theta} = (1 - \theta)\sigma_n + \theta \sigma_{n+1}, \tag{14.10}$$

and

$$\gamma_{n+\theta} = \gamma_{n+\theta}(\sigma_{n+\theta}, \bar{\varepsilon}_{n+\theta}, T_{n+\theta}). \tag{14.11}$$

Within this procedure there are a number of options available depending on the value of the parameter θ. These cases will not be analysed in detail. However, much attention has to be paid to the time step increment strategy as this governs efficiency and accuracy. The time step is always kept within certain limits. If the

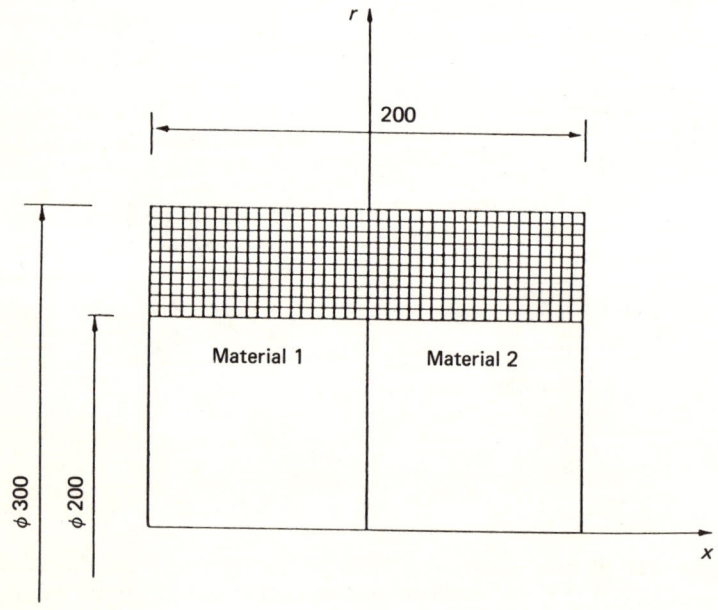

Figure 14.1. Finite-element model of turbine rotor-disc.

time step is too large, accuracy suffers and instability may occur, and if the time step is too small the expense of solving a problem may be very high. The criterion for the time step used for solution is discussed in detail by Zienkiewicz and Cormeau (1974), for instance.

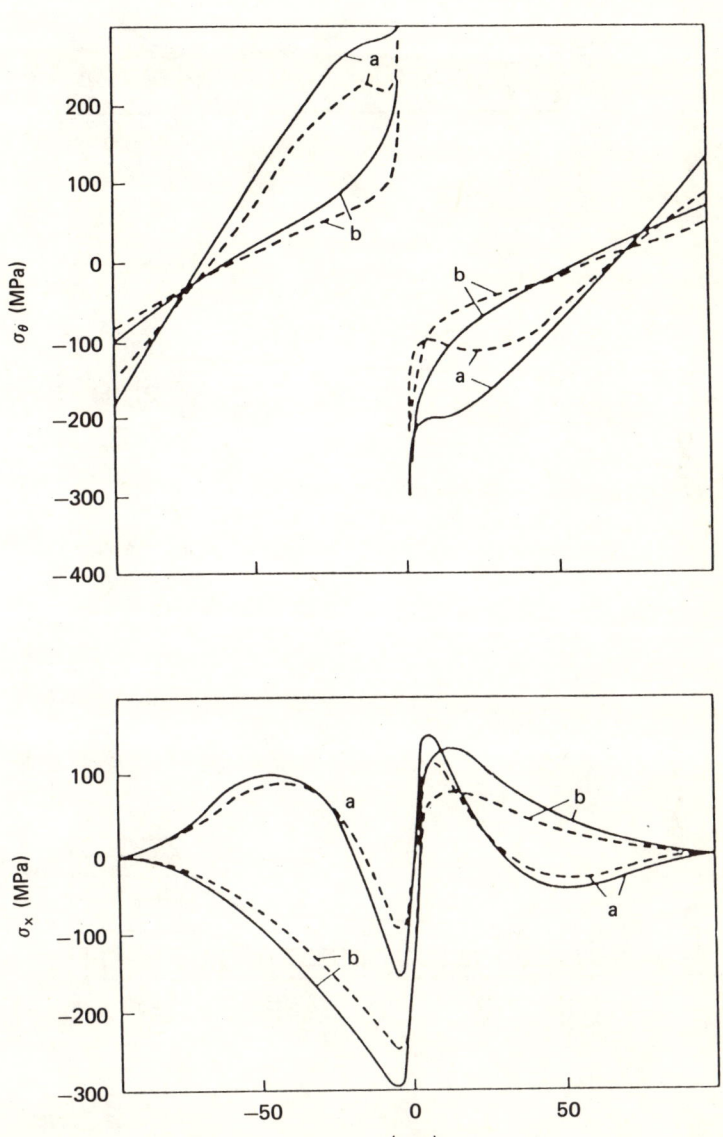

Figure 14.2. Stresses in turbine disc made of two materials.

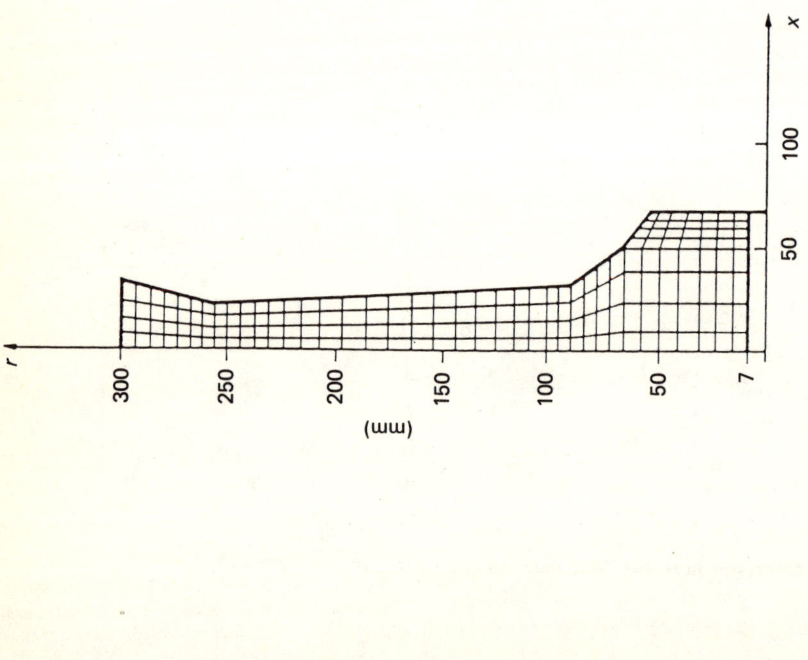

138 INTRODUCTION TO NONLINEAR THERMOMECHANICS

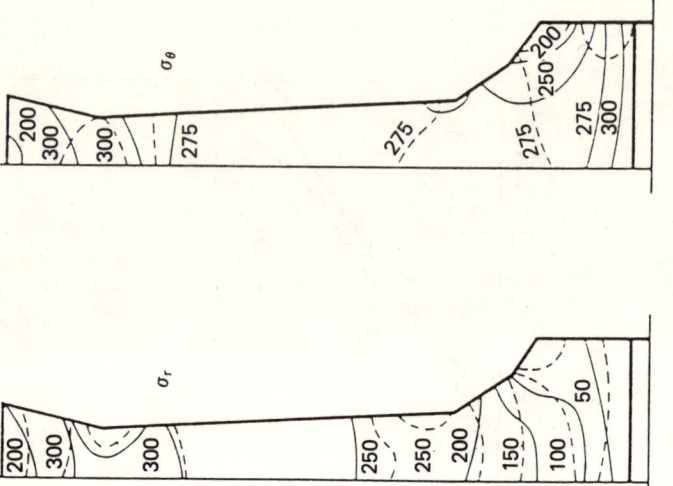

Figure 14.4. Fields of stresses within turbine rotor-disc.

Figure 14.3. Finite-element mesh of a turbine disc.

14.3 Examples

In industry various machine elements consisting of different materials are encountered. The creep problems within such elements are very complicated. Consider the turbine disc shown in Fig. 14.1 (Służalec 1986b). This disc is made of two different materials. The finite-element model is shown in Fig. 14.1. The disc is heated to the temperature 800 °C. It is assumed that $\varepsilon^c = \alpha t^\gamma (e^{\beta\sigma} - 1)$ for $\sigma < \sigma_0$ and $\varepsilon^c = (\sigma/\lambda) - (\sigma_0/\lambda) + \alpha t^\gamma (e^{\beta\sigma} - 1)$ for $\sigma \geq \sigma_0$. The material properties of the first tube at this temperature are $\rho = 7 \times 10^3$ kg m^{-3}, $E = 200$ GPa, Poisson's ratio 0.3, yield stress $\sigma_0 = 35 \times 10^7$ Pa, thermal expansion coefficient 0.12×10^{-4} K^{-1}, $\lambda = 2.16 \times 10^8$ Pa, $n = 4.5$, $\alpha = 0.65 \times 10^{-4}$, $\gamma = 0.5$, $\beta = 1.02 \times 10^{-7}$ m^2 N^{-1}. The second material has the following properties: $\rho = 8 \times 10^3$ kg m^{-3}, $E = 100$ GPa, Poisson's ratio 0.3, yield stress $\sigma_0 = 2 \times 10^8$ Pa, thermal expansion coefficient 0.165×10^{-4} K^{-1}, $\lambda = 4.1 \times 10^8$ Pa, $n = 4.5$, $\alpha = 0.9 \times 10^{-4}$, $\gamma = 0.7$, $\beta = 1.5 \times 10^{-7}$ m^2 N^{-1}. The creep damage relations will not be analysed. Stress values are presented in Fig. 14.2. Stresses at the centres of elements lying on the external surface (a), and on the internal surface of the tube (b), are shown. Continuous lines represent time $t = 0$ and interrupted lines

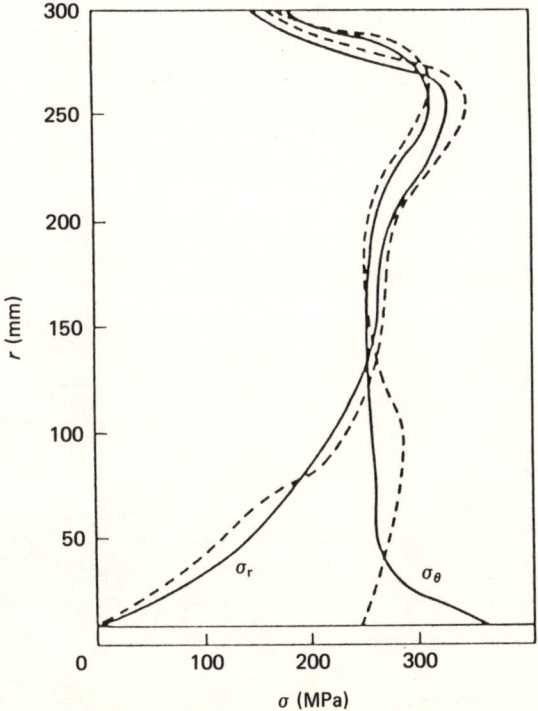

Figure 14.5. Radial and circumferential stresses on the line of axis r.

$t = 250$ hours. The character of these graphs is in agreement with results obtained by Malinin (1975) and Podgornyj (1984).

The next example is the turbine rotor-disc shown in Fig. 14.3, considered for its creep response (Służalec 1986b). The finite-element model is presented in Fig. 14.3, where x is the axis of rotation. The disc is made of steel, is heated to

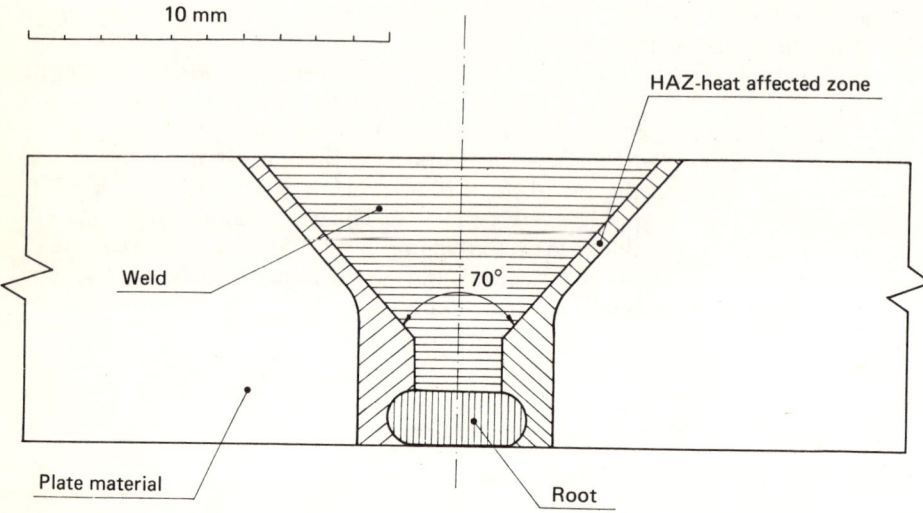

Figure 14.6. Diagram of a V-weld.

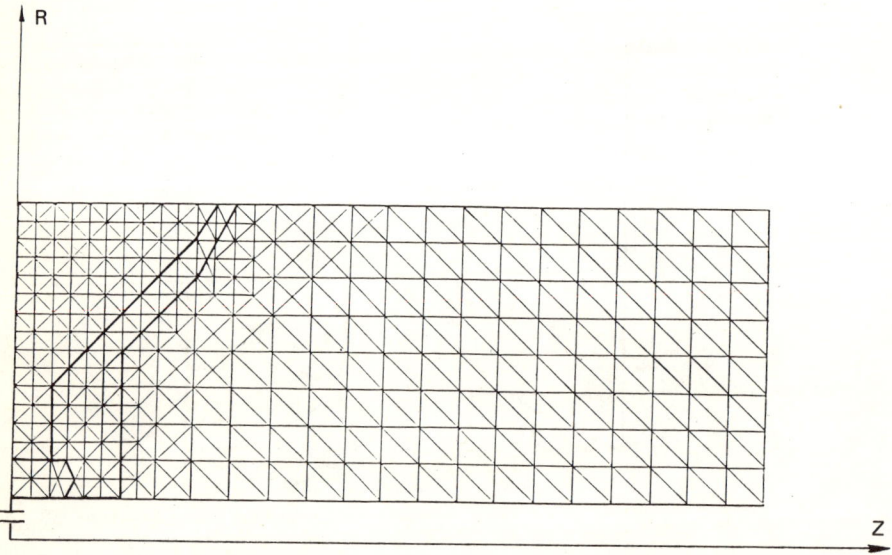

Figure 14.7. Finite-element model of the weld.

FINITE-ELEMENT FORMULATION 141

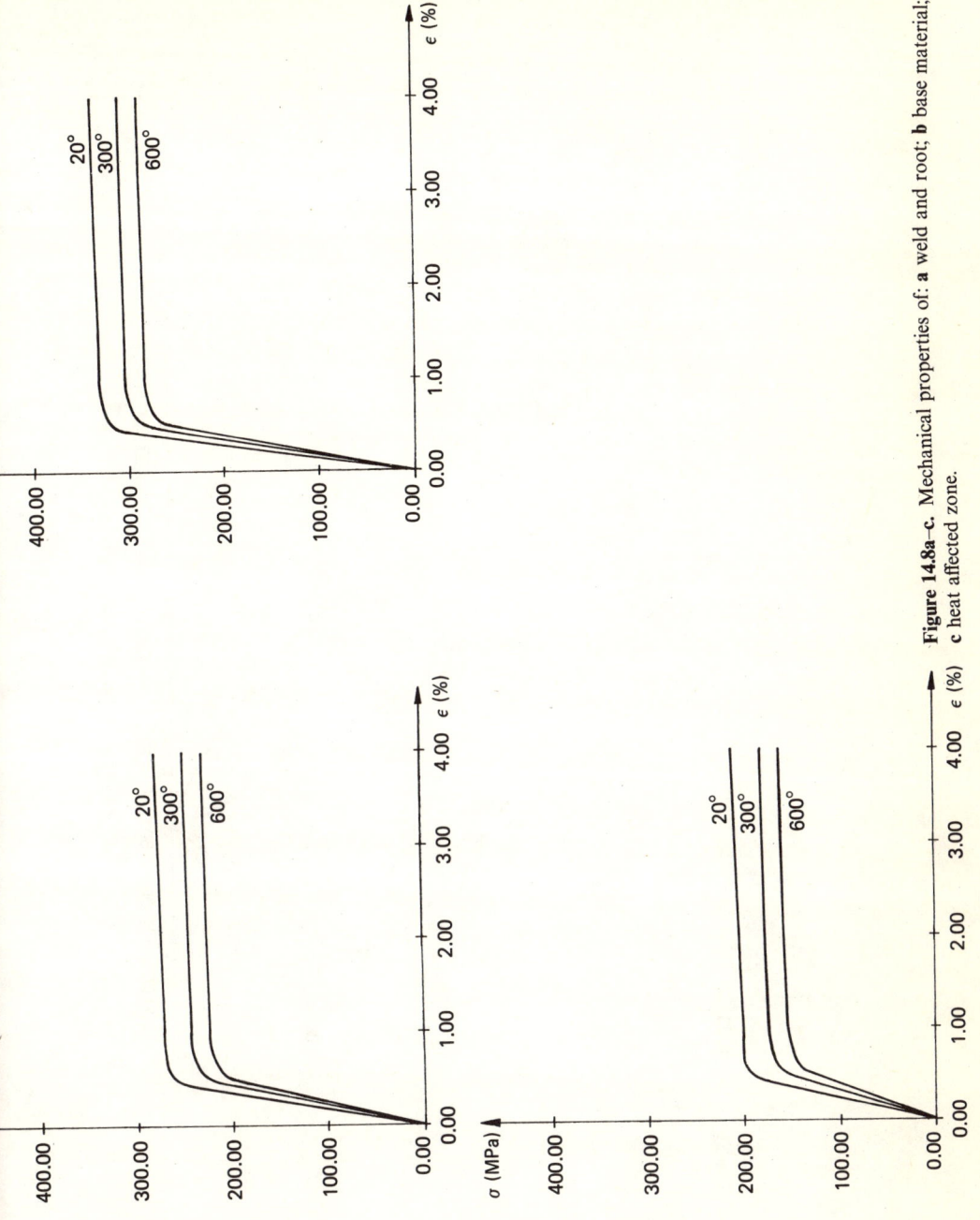

Figure 14.8a–c. Mechanical properties of: **a** weld and root; **b** base material; **c** heat affected zone.

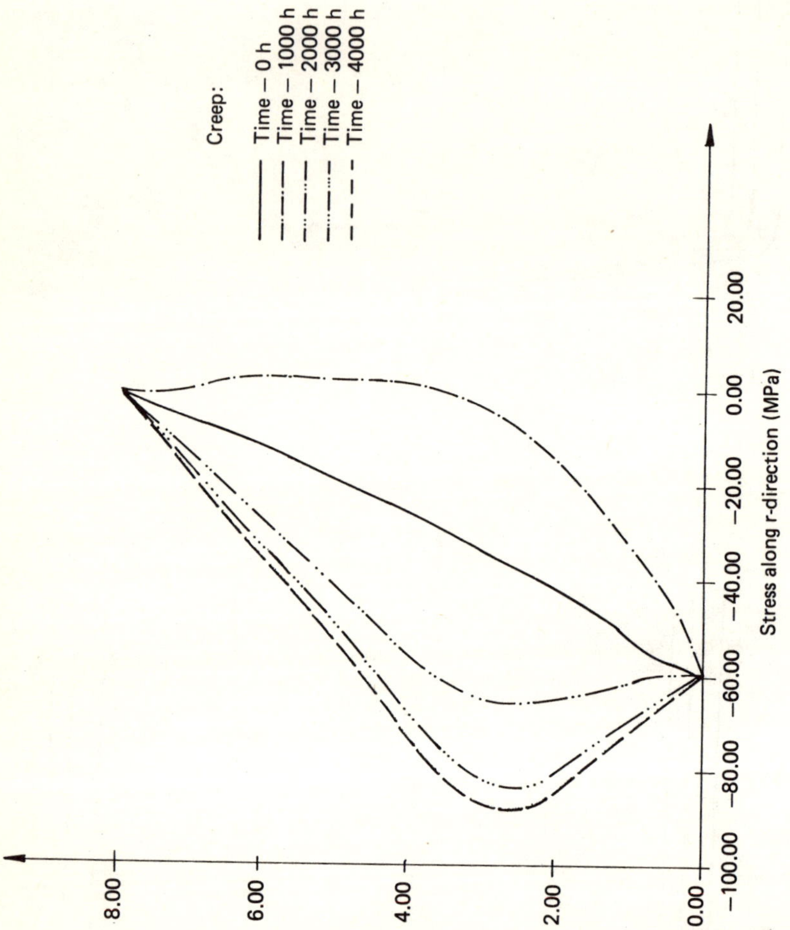
Creep:
Time – 0 h
Time – 1000 h
Time – 2000 h
Time – 3000 h
Time – 4000 h
Thickness (mm)
8.00
6.00
4.00
2.00
0.00
−100.00
−80.00
−60.00
−40.00
−20.00
0.00
20.00
Stress along r-direction (MPa)

a

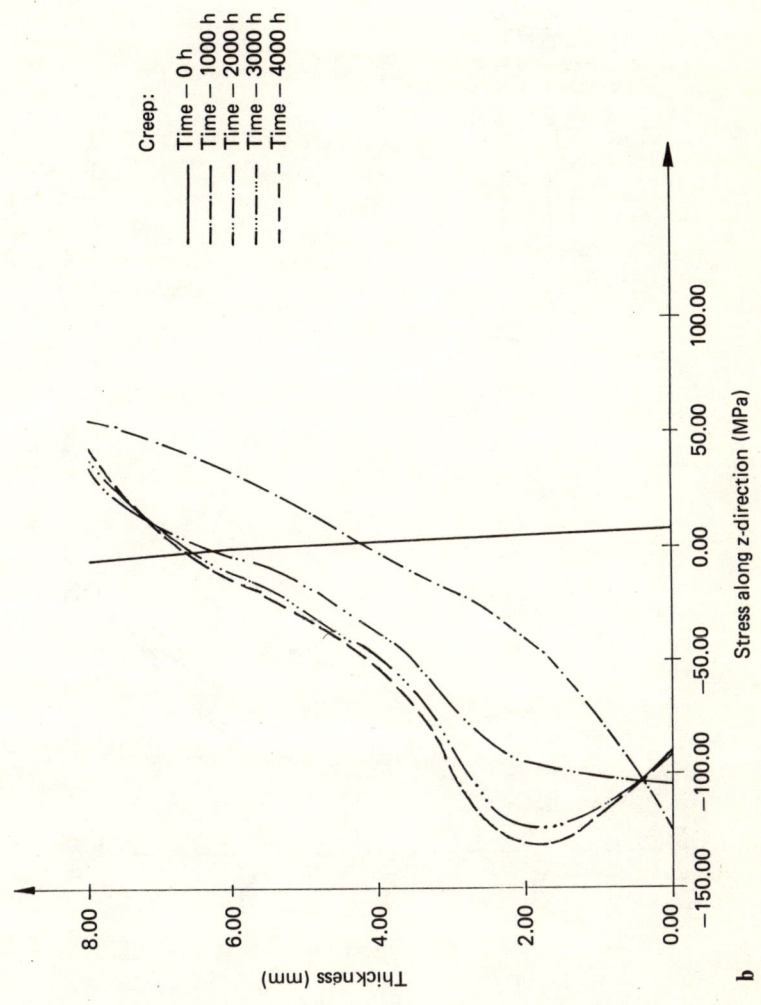

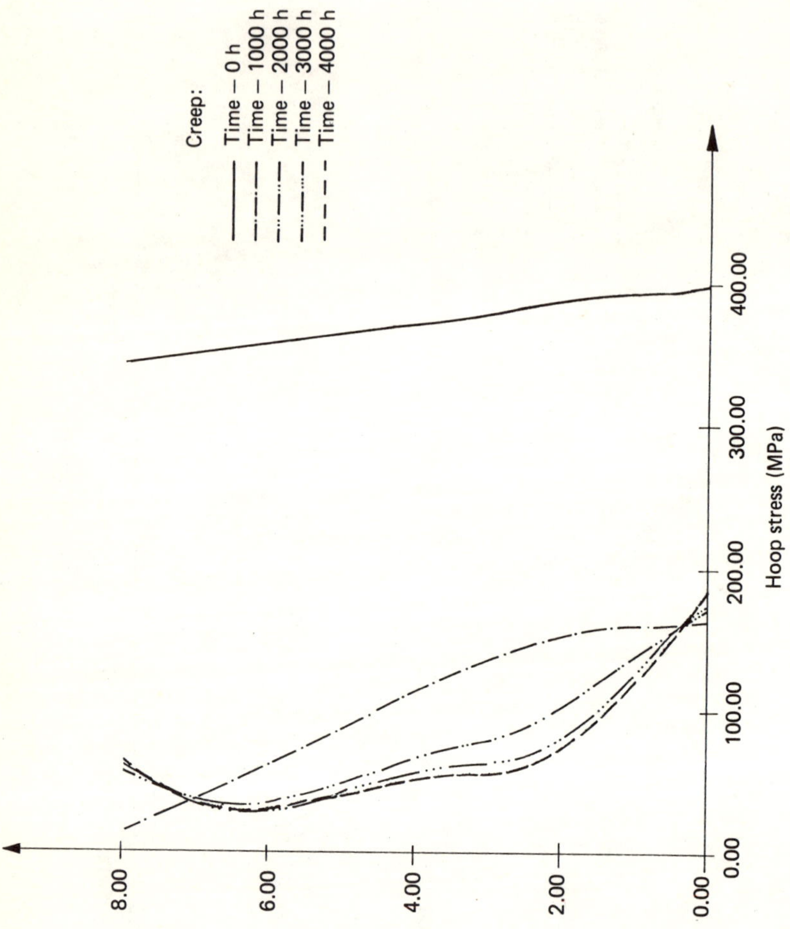

Figure 14.9a–c. Distribution of stresses on the line of weld symmetry: **a** radial stress; **b** axial stress; **c** hoop stress.

FINITE-ELEMENT FORMULATION 145

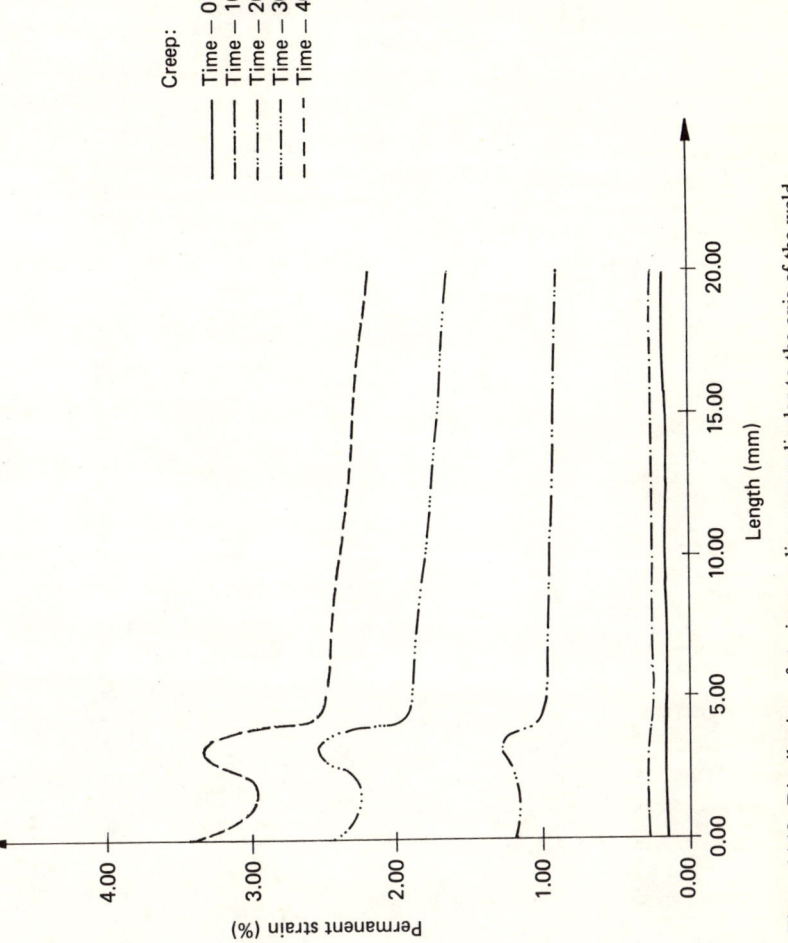

Figure 14.10. Distribution of strain on a line perpendicular to the axis of the weld.

146 INTRODUCTION TO NONLINEAR THERMOMECHANICS

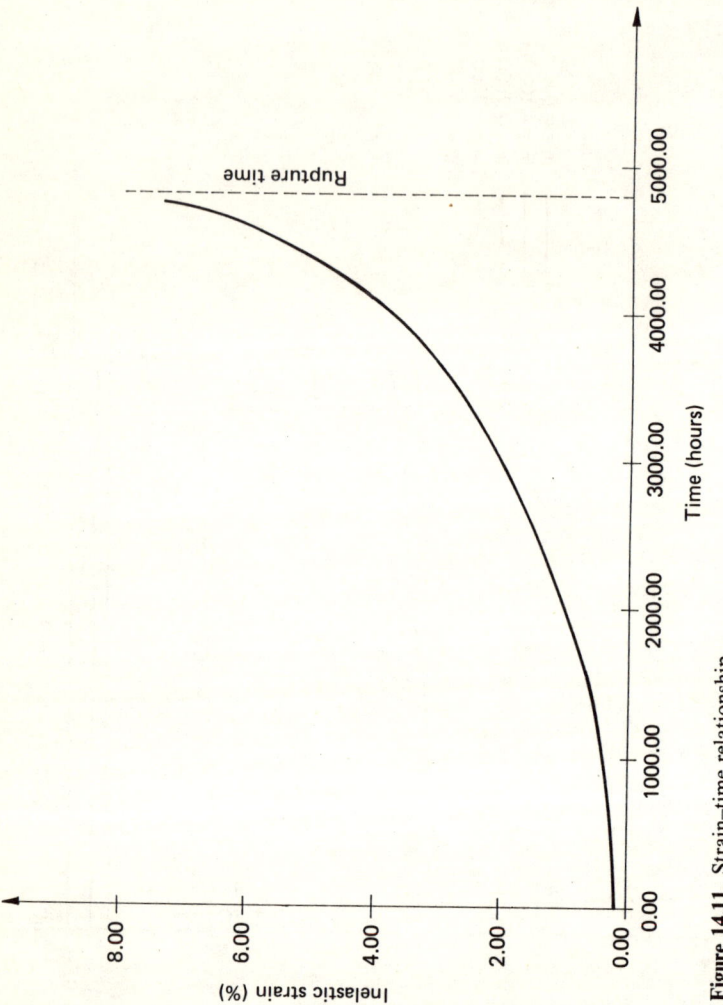

Figure 14.11. Strain–time relationship.

Table 14.1 Creep parameters for all zones of the weld

Zone	n	$K_c [\text{N}^{-n} \text{mm}^{2n} \text{s}^{-1}]$
Weld	2.74	6.5×10^{-8}
Heat affected zone	3.3	3.35×10^{-9}
Root	2.7	6.56×10^{-8}
Base material	3.8	3.54×10^{-17}

600 °C, and rotates with velocity 600 rps. On the surface of the disc the radial forces are assumed to be $P_r = 173$ MPa. The material properties of steel at 600 °C are: $\rho = 0.8 \times 10^4$ kg m^{-3}, $E = 140$ GPa, Poisson's ratio 0.36, yield stress 35×10^7 Pa, thermal expansion coefficient 0 K^{-1}, $\alpha_h = 0$, $C = 3.73 \times 10^{-17}$ m^2 N^{-1} h^{-1}, $v = 2.25$. The radial and circumferential stresses are shown in Figs. 14.4 and 14.5, where continuous lines represent time $t = 0$, and interrupted lines $t = 500$ hours. The critical time obtained is $t = 23 \times 10^2$ hours.

The next example is creep of a welded tube subject to internal pressure (Służalec & Kysiak 1991). The tube is made of 10H2M stainless steel. The metallurgical zones in a V-groove weld are illustrated in Fig. 14.6 and its finite-element model in Fig. 14.7. The thickness of the analysed tube is 8 mm and its diameter is 50 mm. The mechanical properties for all zones, determined by tensile tests, are presented in Fig. 14.8. It is assumed that at 600 °C the creep is characterized by Norton constants (Eq. 11.12). These constants are given in Table 14.1.

We assume that the internal pressure in the tube is 60 MPa and the initial temperature of the tube 20 °C. Non-uniform creep in the zones of the weld causes a stress redistribution. The distribution of stresses on the line of weld symmetry is presented in Fig. 14.9. In Fig. 14.10 we present the distribution of the strain at mid-weld for chosen creep times. The results of calculations for the strain–time relationship are illustrated in Fig. 14.11.

PART V
FINITE STRAINS

15 Finite Strain Models

Characterization of elasto-plastic material behaviour in finite strain regimes is a field of active research where several fundamental questions are yet to be answered. Mathematical consideration lacks a reliable experimental basis and solution is necessarily founded on physical intuition.

Since 1969, when Lee postulated multiplicative decomposition of the gradient of deformation, a considerable body of literature has been devoted to clarifying notions of an intermediate configuration, material frame invariance in conjunction with deformation gradient decomposition, decomposition of strain rates and related aspects (see Nemat-Nasser 1982, 1983; Loret 1983; Dashner 1986). Nevertheless, it is widely recognized that a sound phenomenological theory must rely on micromechanical consideration, as emphasised by Dashner. Significant progress has already been made in modelling single crystals, exploiting the early works of Hill (1978), Rice (1975) and others. Under these circumstances, the computational scene has been faced with a series of unexpected and physically unacceptable results. Unexpected phenomena observed for the kinematic hardening of elasto-plastic materials identified the need for reconstruction of the constitutive theory of finite strain elasto-plasticity, particularly for anisotropic materials, and brought into focus notions of plastic spin and plastically-induced rotation, directors, etc. (see Lee et al. 1983; Nemat-Nasser 1983; Dafalias 1985; Reed and Atluri 1985; Dashner 1986). In computational circles, effort has been directed at the formulation of an algorithm for integration of equations involving constitutive assumptions that would maintain incremental objectivity. Various schemes have been proposed, e.g. Hughes and Winget (1980), Hughes (1984) and Rubinstein and Atluri (1983). Some of these use the rotationally invariant part of the deformation gradient in the hypoelastic constitutive equation and employ logarithmic stretches as strain measures (Nagtegaal and Veldpaus 1984; Rolph and Bathe 1984; Bathe et al. 1986). Since models of finite strain thermo-elasto-plasticity are not yet generally accepted we reduce our consideration to a brief description of the models.

16 Constitutive Equations

16.1 Non-Isothermal Plastic Flow

It is present practice to formulate constitutive equations using assumptions which are not universally accepted, but are admissible in numerical models. Thus, the total Green–Lagrange strain rate tensor in a thermo-elasto-plastic process is expressed as

$$\dot{E}_{IJ} = \dot{E}_{IJ}^e + \dot{E}_{IJ}^T + \dot{E}_{IJ}^{e,T} + \dot{E}_{IJ}^p \tag{16.1}$$

in which $\dot{E}_{IJ}^e$, $\dot{E}_{IJ}^T$, $\dot{E}_{IJ}^{e,T}$ and $\dot{E}_{IJ}^p$ are respectively components of elastic, thermal, temperature-dependent material properties and plastic strain rate tensors. The plastic potential function for the finite strain case takes the form

$$F = F(S_{IJ}, K, T) \tag{16.2}$$

where S_{IJ} are the components of the second Piola–Kirchhoff stress tensor.

If we differentiate F in (16.2) using the chain rule for partial differentiation we obtain

$$\dot{F} = \frac{\partial F}{\partial S_{IJ}} \dot{S}_{IJ} + \frac{\partial F}{\partial K} \dot{K} + \frac{\partial F}{\partial T} \dot{T}. \tag{16.3}$$

If the hardening parameter K is considered to be $dK = S_{IJ} dE_{IJ}^p$ the middle term in (16.3) can be expressed in terms of plastic strains E_{IJ}^p as

$$\frac{\partial F}{\partial K} \dot{K} = \frac{\partial F}{\partial K} \frac{\partial K}{\partial E_{IJ}^p} \dot{E}_{IJ}^p. \tag{16.4}$$

Substituting the above into (16.3) gives

$$\dot{F} = \frac{\partial F}{\partial S_{IJ}} \dot{S}_{IJ} + \frac{\partial F}{\partial K} \frac{\partial K}{\partial E_{IJ}^p} \dot{E}_{IJ}^p + \frac{\partial F}{\partial T} \dot{T}. \tag{16.5}$$

Since equilibrium conditions require that the plastic energy variation be stationary ($\dot{F} = 0$) we get

$$\frac{\partial F}{\partial S_{IJ}} \dot{S}_{IJ} + \frac{\partial F}{\partial K} \frac{\partial K}{\partial E_{IJ}^p} \dot{E}_{IJ}^p + \frac{\partial F}{\partial T} \dot{T} = 0. \tag{16.6}$$

After rearranging (16.1) we can obtain the equations for the elastic strain rate tensor components

$$\dot{E}^e_{IJ} = \dot{E}_{IJ} - \dot{E}^T_{IJ} - \dot{E}^{e,T}_{IJ} - \dot{E}^p_{IJ}. \tag{16.7}$$

Making use of Hooke's law for elastic stress and strain, the second Piola–Kirchhoff stress rate is given as

$$\dot{S}_{IJ} = C^e_{IJKL}(\dot{E}_{KL} - \dot{E}^T_{KL} - \dot{E}^{e,T}_{KL} - \dot{E}^p_{KL}). \tag{16.8}$$

Combining (16.7) and (16.8), and assuming that

$$\dot{E}^p_{IJ} = \lambda \frac{\partial F}{\partial S_{IJ}} \tag{16.9}$$

we get

$$\frac{\partial F}{\partial S_{IJ}} C^e_{IJKL}(\dot{E}_{KL} - \dot{E}^T_{KL} - \dot{E}^{e,T}_{KL} - \dot{E}^p_{KL}) + \frac{\partial F}{\partial K}\frac{\partial K}{\partial E^p_{IJ}}\frac{\partial F}{\partial S_{IJ}} \lambda + \frac{\partial F}{\partial T}\dot{T} = 0. \tag{16.10}$$

After some calculations we obtain the relation for the proportionality factor λ

$$\lambda = \frac{\dfrac{\partial F}{\partial S_{IJ}} C^e_{IJKL}(\dot{E}_{KL} - \dot{E}^T_{KL} - \dot{E}^{e,T}_{KL}) + \dfrac{\partial F}{\partial T}\dot{T}}{\dfrac{\partial F}{\partial S_{PQ}} C^e_{PQRS} \dfrac{\partial F}{\partial S_{RS}} - \dfrac{\partial F}{\partial K}\dfrac{\partial F}{\partial E^p_{PQ}}\dfrac{\partial F}{\partial S_{PQ}}}. \tag{16.11}$$

If we introduce the notation

$$S = \frac{\partial F}{\partial S_{PQ}} C^e_{PQRS} \frac{\partial F}{\partial S_{RS}} - \frac{\partial F}{\partial K}\frac{\partial K}{\partial K^p_{PQ}}\frac{\partial F}{\partial S_{PQ}} \tag{16.12}$$

and

$$\dot{E}^T_{IJ} = \alpha_{IJ}\dot{T}, \tag{16.13}$$

$$D^T_{IJ} = \frac{\partial E_{IJKL}}{\partial T} S_{KL}, \tag{16.14}$$

$$S_{IJ} = E_{IJKL} E_{KL}, \tag{16.15}$$

then from (16.11) and (16.12) we get

$$\lambda = \frac{1}{S}\frac{\partial F}{\partial S_{IJ}} C^e_{IJKL}(\dot{E}_{KL} - \alpha_{KL}\dot{T} - D^T_{KL}\dot{T}) + \frac{\partial F}{\partial T}\dot{T}. \tag{16.16}$$

Next combining (16.8) and (16.16) we have

$$\dot{S}_{IJ} = C^e_{IJKL}\dot{E}_{KL} - C^e_{IJKL}(\alpha_{KL}\dot{T} + D^T_{KL}\dot{T}) - \frac{C^e_{IJVW}}{S}\frac{\partial F}{\partial S_{VW}}$$
$$\times \left[\frac{\partial F}{\partial S_{TU}} C^e_{TUKL}(\dot{E}_{KL} - \alpha_{KL}\dot{T} - D^T_{KL}\dot{T}) + \frac{\partial F}{\partial T}\dot{T}\right]. \tag{16.17}$$

And, rearranging the terms,

$$\dot{S}_{IJ} = C^e_{IJKL}\dot{E}_{KL} - C^e_{IJKL}(\alpha_{KL}\dot{T} + D^T_{KL}\dot{T})$$
$$- \frac{1}{S} C^e_{IJVW} \frac{\partial F}{\partial S_{VW}} \frac{\partial F}{\partial S_{TU}} C^e_{TUKL} \dot{E}_{KL}$$
$$+ \frac{1}{S} C^e_{IJVW} \frac{\partial F}{\partial S_{VW}} \frac{\partial F}{\partial S_{TU}} C^e_{TUKL} (\alpha_{KL}\dot{T} + D^T_{KL}\dot{T}) \quad (16.18)$$
$$- \frac{1}{S} C^e_{IJKL} \frac{\partial F}{\partial S_{KL}} \frac{\partial F}{\partial T} \dot{T}.$$

We can define the plasticity tensor as

$$C^p_{IJKL} = \frac{1}{S} C^e_{IJVW} \frac{\partial F}{\partial S_{VW}} \frac{\partial F}{\partial S_{TU}} C^e_{TUKL} = \frac{1}{S} C^e_{IJVW} S'_{VW} S'_{TU} C^e_{TUKL} \quad (16.19)$$

where S'_{VW} are deviatoric components of the second Piola–Kirchhoff stress tensor. By substituting C^p_{IJKL} as defined above into (16.17) we get

$$\dot{S}_{IJ} = C^e_{IJKL}\dot{E}_{KL} - C^e_{IJKL}(\alpha_{KL}\dot{T} + D^T_{KL}\dot{T}) - C^p_{IJKL}\dot{E}_{KL}$$
$$+ C^p_{IJKL}(\alpha_{KL}\dot{T} + D^T_{KL}\dot{T}) - \frac{1}{S} C^e_{IJKL} \frac{\partial F}{\partial S_{KL}} \frac{\partial F}{\partial T} \dot{T}. \quad (16.20)$$

Defining the elasto-plasticity tensor by

$$C^{ep}_{IJKL} = C^e_{IJKL} - C^p_{IJKL} \quad (16.21)$$

the constitutive equation for non-isothermal plastic flow can be expressed in terms of finite strains as

$$\dot{S}_{IJ} = C^{ep}_{IJKL}\dot{E}_{KL} - C^{ep}_{IJKL}(\alpha_{KL}\dot{T} + D^T_{KL}\dot{T}) - \frac{C^e_{IJKL} S'_{KL}}{S} \frac{\partial F}{\partial T} \dot{T}. \quad (16.22)$$

From (16.19) and using the expression $C^e_{IJKL} S'_{KL} = 2GS'_{IJ}$ we can obtain the relation for C^p_{IJKL} as

$$C^p_{IJKL} = \frac{1}{S} 4G^2 S'_{IJ} S'_{KL}. \quad (16.23)$$

Assume the HM condition, then evaluating the quantity S defined in (16.12) leads to

$$\frac{\partial F}{\partial S_{PQ}} C^e_{PQRS} \frac{\partial F}{\partial S_{RS}} = S'_{PQ} C^e_{PQRS} S'_{RS} = S'_{PQ} 2GS'_{PQ} = \tfrac{4}{3} G\bar{S}^2 \quad (16.24)$$

where

$$\bar{S}^2 = \tfrac{3}{2} S'_{PQ} S'_{PQ}. \quad (16.25)$$

Let H' denote the plastic modulus of the material in a multiaxial stress state:

$$H' = \frac{d\bar{S}}{d\bar{E}^p}. \quad (16.26)$$

Using the HM plastic potential function, we find

$$\frac{\partial F}{\partial K} = \frac{\partial}{\partial K}(J_{2S} - \tfrac{1}{3}\bar{S}^2) = -\tfrac{2}{3}\bar{S}\frac{\partial \bar{S}}{\partial K}. \tag{16.27}$$

Since

$$\frac{d\bar{S}}{dK} = \frac{d\bar{S}}{d\bar{E}^p}\frac{d\bar{E}^p}{dK} = \frac{H'}{\bar{S}} \tag{16.28}$$

we obtain

$$\frac{\partial F}{\partial K} = -\tfrac{2}{3}\bar{S}\frac{H'}{\bar{S}} = -\tfrac{2}{3}H'. \tag{16.29}$$

Hence

$$\frac{\partial F}{\partial K}\frac{\partial K}{\partial E^p_{PQ}}\frac{\partial F}{\partial S_{PQ}} = -\tfrac{2}{3}H'S'_{PQ}S'_{PQ} = -\tfrac{2}{3}H'\tfrac{2}{3}\bar{S}^2 = -\tfrac{4}{9}\bar{S}^2 H' \tag{16.30}$$

and the parameter S can be expressed in terms of the material properties and the state of stress as follows

$$S = \tfrac{4}{3}G\bar{S}^2 - (-\tfrac{4}{9}\bar{S}^2 H') = \tfrac{4}{9}\bar{S}^2(3G + H'). \tag{16.31}$$

As we can see, the basic formulations presented in this section are similar to those presented in Chapter 9, but without the effect of strain-rate-dependent material properties. The general form of the plasticity tensor C^p_{IJKL} is the same as presented in Chapter 9 except for the substitution of s_{ij} by S'_{IJ}. The equivalent stiffness H' here takes a slightly different form

$$H' = \frac{\partial \bar{S}}{\partial \bar{E}^p} \tag{16.32}$$

which requires the definition and availability of the equivalent of S–E curves.

16.2 Multiplicative Decomposition of the Deformation Gradient

In 1990 Wriggers et al. proposed a multiplicative decomposition of the deformation gradient in finite strain thermo-elasto-plasticity of the form

$$F = F_e F_p F_T, \tag{16.33}$$

where the subscripts e, p, and T denote the elastic, plastic and thermal processes respectively. Introducing the multiplicative decomposition of the deformation gradient F into the volume preserving $\hat{F}$ and dilatational part J we get

$$F = J^{1/3}\hat{F}, \tag{16.34}$$

$$\det \hat{F} = 1. \tag{16.35}$$

The classical assumptions of plastic incompressibility and pure dilatational thermal deformation ($J_p = 1$, $\hat{F}_T = 1$) lead to the expressions

$$J = J_e J_T, \tag{16.36}$$

$$\hat{F} = \hat{F}_e \hat{F}_p, \tag{16.37}$$

where

$$F_e = J_e^{1/3}\, \hat{F}_e, \tag{16.38}$$

$$F_p = \hat{F}_p, \tag{16.39}$$

$$F_T = J_T^{1/3}\, \mathbf{1}. \tag{16.40}$$

The plastic strain tensor e^p is introduced by

$$e^p = e - e^e = \tfrac{1}{2}(B_e^{-1} - B^{-1}). \tag{16.41}$$

The rate of deformation tensor d is obtained by the Lie derivative of the total strain e

$$d = L_v(e). \tag{16.42}$$

The thermo-elastic constitutive law is expressed as

$$S = K \ln J_e\, g^{-1} + \mu\, \mathrm{dev}\, \hat{B}_e, \tag{16.43}$$

where S is the Kirchhoff stress, K is the bulk modulus, μ is the shear modulus and

$$J_e = J e^{-3\alpha T}. \tag{16.44}$$

The assumption of a constant temperature, made within the mechanical solution phase, leads to an associated flow rule

$$L_v(e^p) = \dot{\lambda}\frac{\partial F}{\partial S} = \dot{\lambda} n, \tag{16.45}$$

where

$$n = \frac{\mathrm{dev}\, S}{\|\mathrm{dev}\, S\|}. \tag{16.46}$$

The loading conditions may be recast in Kuhn–Tucker form as

$$\dot{\lambda} \geq 0, \quad F \leq 0, \quad \dot{\lambda} F = 0. \tag{16.47}$$

Those readers interested in variational formulation, finite-element equations and other theoretical aspects presented in this section can acquaint themselves with details given by Wriggers et al. (1990).

17 Finite-Element Formulation for Non-Isothermal Plastic Flow

17.1 Total Lagrange Formulation (TL)

The constitutive law may be used in conjunction with the law of conservation of momentum by way of an appropriate variational principle.

In a TL description all variables in an equation of motion considered at time $t + \Delta t$ are referred to the configuration at time $t = 0$. Consider the virtual work formulation

$$\int_{_0V} {}_0S_{IJ}^{n+1}\, \delta_0 E_{IJ}^{n+1}\, d_0V = \int_{_0S} {}_0t_I^{n+1}\, \delta_0 u_I\, d_0S + \int_{_0V} {}_0\rho_0 f_I^{n+1}\, \delta_0 u_I\, d_0V \qquad (17.1)$$

where ${}_0S_{IJ}^{n+1}$ is the second Piola–Kirchhoff stress tensor at time $t + \Delta t$ referred to the initial configuration at time $t = 0$, $\delta_0 E_{IJ}^{n+1}$ is the variation of the Green–Lagrange strains at $t + \Delta t$ referred to the initial configuration ${}_0V$, ${}_0t_I^{n+1}$ is the surface traction at time $t + \Delta t$ referred to the surface of the configuration ${}_0S$, $\delta_0 u_I$ is the variation of displacement, ${}_0\rho$ is the local density in the initial configuration and ${}_0f_I^{n+1}$ is the body force per unit mass at time $t + \Delta t$ referred to the initial configuration ${}_0V$.

The virtual work equation is incrementalized using

$${}_0S_{IJ}^{n+1} = {}_0S_{IJ}^n + \Delta_0 S_{IJ}, \qquad (17.2)$$

and

$${}_0E_{IJ}^{n+1} = {}_0E_{IJ}^n + \Delta_0 E_{IJ}, \qquad (17.3)$$

where $\Delta_0 S_{IJ}$ and $\Delta_0 E_{IJ}$ are increments of the second Piola–Kirchhoff stress and the Green–Lagrange strain tensor components, respectively.

The increment of the Green–Lagrange strain tensor can be written in the form

$$\Delta_0 E_{IJ} = \Delta_0 \varepsilon_{IJ} + \Delta_0 \eta_{IJ}, \qquad (17.4)$$

where

$$\Delta_0 \varepsilon_{IJ} = \tfrac{1}{2}(\Delta_0 u_{I,J} + \Delta_0 u_{J,I} + {}_0u_{K,I}^n \Delta_0 u_{K,J} + \Delta_0 u_{K,I}\, {}_0u_{K,J}^n), \qquad (17.5)$$

$$\Delta_0 \eta_{IJ} = \tfrac{1}{2} \Delta_0 u_{K,I}\, \Delta_0 u_{K,J}. \qquad (17.6)$$

denote the linear and nonlinear parts, respectively. Write Eq. (16.22) in the form

$$\Delta_0 S_{IJ} = {}_0C^{ep}_{IJKL}\Delta_0 E_{KL} - \Delta_0 S^R_{IJ}, \qquad (17.7)$$

where

$$\Delta_0 S^R_{IJ} = {}_0C^{ep}_{IJKL}({}_0\alpha_{KL}\Delta T + {}_0D^T_{KL}\Delta T) + \Delta_0 P_{IJ}, \qquad (17.8)$$

and

$$\Delta_0 P_{IJ} = \frac{{}_0C^e_{IJKL}S'_{KL}}{{}_0S}\frac{\partial F}{\partial T}\Delta T. \qquad (17.9)$$

After linearization of (17.7) one obtains

$$\Delta_0 S_{IJ} = {}_0C^{ep}_{IJKL}\Delta_0 \varepsilon_{KL} - \Delta_0 S^R_{IJ}. \qquad (17.10)$$

With the help of the latter and

$$\delta\Delta_0 E_{IJ} = \delta\Delta_0 \varepsilon_{IJ} \qquad (17.11)$$

Eq. (17.1) can be written as

$$\int_{0V}\delta\Delta_0\varepsilon_{IJ}\,{}_0C^{ep}_{IJKL}\Delta_0\varepsilon_{KL}d_0V + \int_{0V}\delta\Delta_0\eta_{IJO}S^n_{IJ}d_0V = \int_{0S}{}_0t^{n+1}_I\delta_0u_I d_0 S$$

$$+ \int_{0V}{}_0\rho_0 f^{n+1}_I\delta_0u_I d_0V - \int_{0V}\delta\Delta_0\varepsilon_{IJ}({}_0S^n_{IJ} - \Delta_0 S^R_{IJ})d_0V. \qquad (17.12)$$

The resulting equation in matrix notation is

$$({}_0\mathbf{K}^n_L + {}_0\mathbf{K}^n_{NL})\Delta\mathbf{u} = {}_0\mathbf{R}^{n+1} - {}_0\mathbf{F}^n, \qquad (17.13)$$

where

$$_0\mathbf{K}_L = \int_{0V}{}_0\mathbf{B}^T_L\,{}_0\mathbf{C}^{ep}\,{}_0\mathbf{B}_L d_0 V, \qquad (17.14)$$

$$_0\mathbf{K}_{NL} = \int_{0V}{}_0\mathbf{B}^T_{NL}({}_0\mathbf{S}^n - {}_0\mathbf{C}^{ep}\,{}_0\mathbf{a}\Delta T - {}_0\mathbf{C}^{ep}\,{}_0\mathbf{D}^T\Delta T - \Delta_0\mathbf{P})_0\mathbf{B}_{NL}d_0 V, \qquad (17.15)$$

$$_0R^{n+1}_K = \int_{0S}{}_0t^{n+1}_I\frac{\partial\Delta u_I}{\partial\Delta u_K}d_0 S + \int_{0V}{}_0\rho_0 f^{n+1}_I\frac{\partial\Delta u_I}{\partial\Delta u_K}d_0 V, \qquad (17.16)$$

$$-{}_0\mathbf{F}^n = \int_{0V}{}_0\mathbf{B}^T_L({}_0\mathbf{S}^n - {}_0\mathbf{C}^{ep}\,{}_0\mathbf{a}\Delta T - {}_0\mathbf{C}^{ep}\,{}_0\mathbf{D}^T\Delta T - \Delta_0\mathbf{P})d_0 V, \qquad (17.17)$$

and $\mathbf{B}_L$ and $\mathbf{B}_{NL}$ are the linear strain displacement transformation matrix and the nonlinear strain displacement transformation matrix, respectively.

17.2 Updated Lagrange (UL) and Updated Lagrange–Jaumann Formulations (ULJ)

Assuming that the equation of motion is solved for time step n, we can express the UL as

$$\int_{nV} {}_nS_{IJ}^{n+1} \delta_n E_{IJ}^{n+1} dV = \int_{nS} {}_n t_I^{n+1} \delta_n u_I d_n S + \int_{nV} {}_n\rho_n f_I^{n+1} \delta_n u_I d_n V \qquad (17.18)$$

where ${}_nS_{IJ}^{n+1}$ is the second Piola–Kirchhoff stress at time $t + \Delta t$ referred to the configuration at time t, ${}_nE_{IJ}^{n+1}$ is the Green–Lagrange strain tensor at time $t + \Delta t$ referred to the configuration at time t, ${}_n\rho$ is the density in the configuration at time $t + \Delta t$ referred to the configuration at time t, and ${}_nt_I^{n+1}$ and ${}_nf_I^{n+1}$ are surface tractions and body forces at time $t + \Delta t$ referred to the current configuration at time t.

We make use of the incremental decompositions

$${}_nS_{IJ}^{n+1} = {}_n\sigma_{IJ}^n + \Delta_n S_{IJ}, \qquad (17.19)$$

$${}_nE_{IJ}^{n+1} = {}_nE_{IJ}^n + \Delta_n E_{IJ}, \qquad (17.20)$$

where σ_{IJ}^n is the Cauchy stress tensor, and the fact that the increment of the Green–Lagrange strain tensor can be written as the sum

$$\Delta_n E_{IJ} = \Delta_n \varepsilon_{IJ} + \Delta_n \eta_{IJ}, \qquad (17.21)$$

where

$$\Delta_n \varepsilon_{IJ} = \tfrac{1}{2}(\Delta_n u_{I,J} + \Delta_n u_{J,I}), \qquad (17.22)$$

$$\Delta_n \eta_{IJ} = \tfrac{1}{2} \Delta_n u_{K,I} \Delta_n u_{K,J}. \qquad (17.23)$$

Introducing approximations similar to those in the TL description, we have

$$\Delta_n S_{IJ} = {}_nC_{IJKL}^{ep} \Delta_n \varepsilon_{KL} - \Delta_n \sigma_{IJ}^R, \qquad (17.24)$$

$$\delta \Delta_n E_{IJ} = \delta \Delta_n \varepsilon_{IJ}, \qquad (17.25)$$

and Eq. (17.18) takes the form

$$\int_{nV} \delta\Delta_n\varepsilon_{IJ} {}_nC_{IJKL}^{ep} \Delta_n\varepsilon_{KL} d_nV + \int_{nV} \delta\Delta_n\eta_{IJ} {}_n\sigma_{IJ}^n d_nV = \int_{nS} {}_nt_I^{n+1} \delta_n u_I d_n S$$

$$+ \int_{nV} {}_n\rho_n f_I^{n+1} \delta_n u_I d_n V - \int_{nV} \delta\Delta_n\varepsilon_{IJ}(\sigma_{IJ}^n - \Delta_n\sigma_{IJ}^R) d_nV. \qquad (17.26)$$

The resulting matrix equation is

$$({}_n\mathbf{K}_L^n + {}_n\mathbf{K}_{NL}^n) \Delta\mathbf{u} = {}_n\mathbf{R}^{n+1} - {}_n\mathbf{F}^n, \qquad (17.27)$$

where

$${}_n\mathbf{K}_L = \int_{nV} {}_n\mathbf{B}_L^T {}_n\mathbf{C}^{ep} {}_n\mathbf{B}_L \, d_nV, \qquad (17.28)$$

$${}_n\mathbf{K}_{NL} = \int_{nV} {}_n\mathbf{B}_{NL}^T ({}_n\sigma^n - \mathbf{C}^{ep} {}_n a \Delta T - \mathbf{C}^{ep} {}_n\mathbf{D}^T \Delta T - \Delta_n\mathbf{P}) \mathbf{B}_{NL} d_nV, \qquad (17.29)$$

$${}_n R_K^{n+1} = \int_{nS} {}_nt_I^{n+1} \frac{\partial \Delta u_I}{\partial \Delta u_K} d_nS + \int_{nV} \rho_n f_I^{n+1} \frac{\partial \Delta u_I}{\partial \Delta u_K} d_nV, \qquad (17.30)$$

$$-{}_n\mathbf{F}^n = \int_{nV} {}_n\mathbf{B}_L^T ({}_n\sigma^n - \mathbf{C}^{ep} {}_n a \Delta T - \mathbf{C}^{ep} {}_n\mathbf{D}^T \Delta T - \Delta_n\mathbf{P}) d_nV. \qquad (17.31)$$

In ULJ formulation one assumes the equation (17.26) used for UL formulation and, taking into account conditions of objectivity and conjugacy of the constitutive quantities, one uses for subsequent computations the Jaumann stress increment (for further details see Washizu (1982) for instance)

$$\Delta \sigma_{IJ}^{J} = \Delta \sigma_{IJ} + \sigma_{IP}^{n} \Delta_{n} \omega_{PJ} - \sigma_{JP}^{n} \Delta_{n} \omega_{PI}, \qquad (17.32)$$

where

$$\Delta_{n} \omega_{IJ} = \tfrac{1}{2}(\Delta_{n} u_{J,I} - \Delta_{n} u_{I,J}) \qquad (17.33)$$

is the increment of the spin tensor.

17.3 Updated Lagrange–Hughes Formulation (ULH)

For the ULH formulation one applies Eq. (17.26) with minor changes proposed by Hughes (1980). The algorithm consists of the following steps:

Compute the displacement gradient in the configuration at time step $n + \tfrac{1}{2}$ (time $t + \tfrac{1}{2}\Delta t$)

$$G_{IJ} = \frac{\partial \Delta_{n} u_{I}}{\partial \Delta_{n+\tfrac{1}{2}} x_{J}}. \qquad (17.34)$$

Define objective strain and rotation increments in terms of G_{IJ} as follows

$$\Delta \zeta_{IJ} = \tfrac{1}{2}(G_{IJ} + G_{JI}), \qquad (17.35)$$

$$\Delta \omega_{IJ} = \tfrac{1}{2}(G_{IJ} - G_{JI}). \qquad (17.36)$$

Define the objective Cauchy stress increment as

$$\Delta \sigma_{IJ} = {}_{n+\tfrac{1}{2}}C_{IJKL}^{ep} \Delta \zeta_{KL} - \Delta_{n+\tfrac{1}{2}} \sigma_{KL}^{R}. \qquad (17.37)$$

The final stresses and strains in the configuration at time step $n + \tfrac{1}{2}$ are then given by

$$\sigma_{IJ}^{n+\tfrac{1}{2}} = \tilde{\sigma}_{IJ} + \Delta \sigma_{IJ}, \qquad (17.38)$$

$$\varepsilon_{IJ}^{n+\tfrac{1}{2}} = \tilde{\varepsilon}_{IJ} + \Delta \zeta_{IJ}, \qquad (17.39)$$

where

$$\tilde{\sigma}_{IJ} = Q_{IR} Q_{JS} \sigma_{RS}^{n}, \qquad (17.40)$$

$$\tilde{\varepsilon}_{IJ} = Q_{IR} Q_{JS} \varepsilon_{RS}^{n}, \qquad (17.41)$$

and $Q_{NM} = Q_{HM}(\omega_{NM})$ is a rotation matrix (Hughes 1980).

Then the stress and strain given by (17.38) and (17.39) are rotated from the configuration at time step $n + \tfrac{1}{2}$ to the configuration at step $n + 1$ by the same process as in Eqs. (17.40) and (17.41).

PART VI
COUPLED THERMO-PLASTICITY

18 Equations of Coupled Thermo-Plasticity

The interaction of temperature and deformation during plastic flow appears in various forms. A thermal field influences the material properties, modifies the extent of plastic zones and results in ratchetting during cyclic heating etc., but also the deformation induces changes in the temperature distribution. From the stress analysis point of view the thermal effects in plasticity can be studied at two levels, depending on whether uncoupled or coupled theories of thermomechanical response have to be employed. The most technologically important problems regarding ratchetting, stresses in welding, residual stresses after quenching, design of reactor fuel elements etc. can be satisfactorily studied within an uncoupled theory. In such an approach the temperature enters the stress–strain relation through the thermal dilatation only, and possibly influences the material constants. The heat conduction equation and the relations governing the stress field are considered separately. These problems were described in Sections 9, 10 and 15. There exist, however, many instances when coupling of thermal and deformation states is of importance. We mention here stability analysis for metal forming, catastrophic shear during machining, and fatigue (Służalec 1990b).

18.1 Heat Transfer Equations

Consider a differential quantity

$$(Tq_i)_{,i} = Tq_{i,i} + q_i T_{,i}, \tag{18.1}$$

or

$$Tq_{i,i} = (Tq_i)_{,i} - q_i T_{,i}. \tag{18.2}$$

Integrating the above expression, we obtain

$$\int_V Tq_{i,i} dV = \int_V (Tq_i)_{,i} dV - \int_V q_i T_{,i} dV, \tag{18.3}$$

which, on using the Gauss–Green theorem

$$\int_V (Tq_i)_{,i} dV = \int_S Tq_i n_i dS \tag{18.4}$$

becomes

$$\int_V T q_{i,i} dV = -\int_V q_i T_{,i} dV + \int_S T q_i n_i dS. \tag{18.5}$$

In order to describe the heat transfer in the solid we have to determine a magnitude $q_{i,i}$. The magnitude $q_{i,i}$ can be obtained from expressions (4.29)–(4.32) if the specific free enthalpy $\psi = \psi(\sigma_{ik}, T, b, \beta_{ik})$ is known (see Bruhns and Służalec 1989). Neglecting the internal variables the specific free enthalpy ψ can be expressed as

$$\psi = \psi_0 - \frac{E}{2\rho}\varepsilon^{e2} - \frac{c_v}{2T}(T - T_0)^2, \tag{18.6}$$

where c_v is the specific heat at constant volume and T_0 is the reference temperature. It should be noted that internal variables can be introduced also in the approach presented (Bruhns and Służalec 1989).

From (18.6) we obtain

$$-T\frac{\partial^2 \psi}{\partial T^2} = c_v. \tag{18.7}$$

Finally we get the balance equation for specific reversible work (Eq. (4.30))

$$\dot{w}^{(r)} = \frac{1}{\rho}\sigma_{ik}\overset{v}{\varepsilon}_{ki}^{(r)} = -\sigma_{ik}\left\{\frac{\partial^2 \psi}{\partial \sigma_{rs} \partial \sigma_{ik}}\overset{v}{\sigma}_{rs} + \frac{\partial^2 \psi}{\partial T \partial \sigma_{ik}}\dot{T}\right\}, \tag{18.8}$$

and the balance equation for remaining specific energy supply (Eq. (4.31))

$$\dot{w}^{(i)} - \frac{1}{\rho}q_{i,i} + r = -T\frac{\partial^2 \psi}{\partial \sigma_{ik} \partial T}\overset{v}{\sigma}_{ik} + c_v \dot{T}. \tag{18.9}$$

By Eq. (4.26)

$$\varepsilon_{ik}^e = -\rho\frac{\partial \psi}{\partial \sigma_{ki}}, \tag{18.10}$$

and hence

$$-\rho\frac{\partial^2 \psi}{\partial T \partial \sigma_{ki}} = \frac{\partial \varepsilon_{ik}^e}{\partial T} = \gamma_{ik}. \tag{18.11}$$

By Eqs. (18.9) and (18.11) we get

$$\dot{w}^{(i)} - \frac{1}{\rho}q_{i,i} + r = \frac{1}{\rho}T\gamma_{ik}\overset{v}{\sigma}_{ik} + c_v \dot{T} \tag{18.12}$$

and hence

$$\dot{u}^{(d)} - q_{i,i} + \rho r = T\gamma_{ik}\overset{v}{\sigma}_{ik} + \rho c_v \dot{T} \tag{18.13}$$

where $\dot{u}^{(d)}$ is the so-called internal dissipation function.

By (18.5) and (18.13) we obtain

$$\int_V T(\dot{u}^{(d)} + \rho r - \rho c_v \dot{T} - T\gamma_{ik}\overset{v}{\sigma}_{ik}) dV = -\int_V q_i T_{,i} dV + \int_S T q_i n_i dS. \tag{18.14}$$

It should be noted that if, following common practice, we introduce the tensor β_{ik}

$$\beta_{ik} = \frac{\partial \sigma_{ik}}{\partial T},$$

Eq. (18.14) can be expressed as

$$\int_V T(\rho c_v \dot{T} + T\beta_{ik}\dot{\varepsilon}_{ik} - \dot{u}^{(d)} - \rho r)dV = \int_V q_i T_{,i} dV - \int_S Tq_i n_i dS. \qquad (18.15)$$

18.2 Finite-Element Formulation for the Heat Flow Equation

We recall briefly the fundamental relations used in discretized procedures. The components of the displacement vector in an element U can be expressed by nodal displacements u as

$$U(x, t) = \mathbf{N}(x)u(x, t), \qquad (18.16)$$

which leads to

$$V(x, t) = \mathbf{N}(x)\dot{u}(x, t). \qquad (18.17)$$

For temperature and temperature gradient we assume the following interpolation functions

$$T = \mathbf{H}T(t), \qquad (18.18)$$

$$VT = \mathbf{V}\mathbf{H}T(t). \qquad (18.19)$$

The strain matrix can be expressed as

$$\varepsilon(x, t) = \mathbf{B}u(t). \qquad (18.20)$$

Upon substituting (18.16)–(18.20) into (18.15) one infers that

$$\mathbf{C}\dot{T}(t) + \mathbf{K}_T T(t) + \mathbf{M}_u \dot{u}(t) - D - Q = 0, \qquad (18.21)$$

where

$$\mathbf{C} = \int_V \rho c_v \mathbf{H}\mathbf{H}^T dV \text{ is the capacity matrix}, \qquad (18.22)$$

$$\mathbf{K}_T = \int_V \mathbf{V}\mathbf{H}k\mathbf{V}\mathbf{H}^T dV \text{ is the conductivity matrix}, \qquad (18.23)$$

$$\mathbf{M}_u = \int_V \mathbf{H}\mathbf{B}\beta dV \text{ is the thermomechanical coupling matrix}, \qquad (18.24)$$

$$D = \int_V \dot{u}^{(d)} \mathbf{H} dV \text{ is the dissipation vector} \qquad (18.25)$$

and

$$Q = \int_V \mathbf{H}\rho r \, dV + \int_V \mathbf{H}q^T n \, dV \text{ is the thermal load vector}. \qquad (18.26)$$

It should be recalled, that in a majority of practical engineering problems the thermomechanical coupling matrix $\mathbf{M}_u$ does not play an important role and can be omitted. Then (18.21) reduces to

$$\mathbf{C}\dot{\mathbf{T}}(t) + \mathbf{K}_T \mathbf{T}(t) - \mathbf{D} - \mathbf{Q} = 0. \tag{18.27}$$

18.3 Internal Dissipation Function

The dissipation of internal energy was recognized in 1925, by Farren and Taylor. However a mathematical description of this phenomenon had not been formulated until fairly recently. Dillon (1962b) reported some interesting observations made in experiments on the cyclic twisting of circular bars. He discovered that cyclic dilatation of the specimen alone induced no temperature rise whereas twisting the specimen in alternating clockwise and anti-clockwise directions produced heat. From these experiments, one concludes that deviatoric deformations were the main reason for heat generation in the solid, even in an elastically deforming solid. In a subsequent paper Dillon (1963) identified work due to plastic deformation as the major contributor to the heat effect and proposed a mathematical expression for the internal dissipation rate, namely

$$\dot{u}^{(d)} = \sigma_{ij} \dot{\varepsilon}^p_{ij} = s_{ij} \dot{e}^p_{ij}. \tag{18.28}$$

There is a general acceptance that not all the mechanical work produced by plastic deformation is converted into thermal energy in the solid. A larger portion of that work is believed to have been spent in the change of material microscopic structure. For the evaluation of parameter $\dot{u}^{(d)}$ various factors have been proposed and applied. Lee (1969) proposed a variable factor χ, so that

$$\dot{u}^{(d)} = \chi \sigma_{ij} \dot{\varepsilon}^p_{ij}, \tag{18.29}$$

with χ varying from 0.9 to unity with increasing plastic deformation. A similar expression was proposed by Lehmann (1980) where $\chi < 1.0$ and χ is a function of both temperature T and the work-hardening parameter K of the material.

Nied and Batterman (1972) proposed the following form for the internal dissipation function

$$\dot{u}^{(d)} = (1 - \Lambda)\sigma_{ij} \dot{\varepsilon}^p_{ij}, \tag{18.30}$$

in which the parameter $\Lambda = \Lambda(T)$ is a measure of the ratio of stored energy to plastic energy expended under adiabatic conditions. The numerical value of Λ increases with temperature. It can also be regarded as the ratio of the conversion of the kinetic energy to the internal and adiabatic inelastic deformation.

Raniecki and Sawczuk (1975) and Mroz and Raniecki (1976) proposed a slightly different form for the energy dissipation

$$\dot{u}^{(d)} = \sigma_{ij} \dot{\varepsilon}^p_{ij} - \Pi \dot{K} \tag{18.31}$$

where Π is called the conjugate to the internal state variable of hardening parameter K. Mroz and Raniecki used the following relation for determination of parameter Π

$$\Pi = -\frac{\partial \dot{u}^{(d)}}{\partial K}. \tag{18.32}$$

It is obvious from the above description that an accurate estimate of the magnitude of the internal energy dissipation $\dot{u}^{(d)}$ is possible if the factors χ, Π and Λ are accurately determined. Służalec (1990a) carried out some experiments as to the magnitude of Λ in the case of elasto-plastic tension of a cylindrical specimen. The value obtained was 0.85. In general, a range of 0.1 to 0.9 is used for the parameter Λ depending on the kind of thermomechanical process.

18.4 Stress–Strain Relations in Coupled Thermo-Plasticity

18.4.1 Thermo-Elasto-Plastic Model Based on Additive Decomposition of Strain

Consider the basic equation in thermo-elasto-plasticity, i.e. Eq. (9.15) without the strain rate effect, or Eq. (16.22) for finite strains

$$\dot{\sigma}_{ij} = C^{ep}_{ijkl}\dot{\varepsilon}_{kl} - C^{ep}_{ijkl}\alpha_{kl}\dot{T} - C^{ep}_{ijkl}D^T_{kl}\dot{T} - \frac{C^e_{ijkl}s_{kl}}{S}\frac{\partial F}{\partial T}\dot{T}. \tag{18.33}$$

If we introduce the tensor

$$\zeta_{ij} = -C^{ep}_{ijkl}(\alpha_{kl} + D^T_{kl}) - \frac{C^e_{ijkl}s_{kl}}{S}\frac{\partial F}{\partial T}, \tag{18.34}$$

then we can write Eq. (18.33) in the form

$$\dot{\sigma}_{ij} = C^{ep}_{ijkl}\dot{\varepsilon}_{kl} + \zeta_{ij}\dot{T}. \tag{18.35}$$

The universal equilibrium condition for a solid subject to mechanical loads is

$$\int_V \dot{\sigma}_{ij}\delta\dot{\varepsilon}_{ij}\,dV = \int_V \dot{f}_i\delta\dot{u}_i\,dV + \int_S \dot{t}_i\delta\dot{u}_i\,dS, \tag{18.36}$$

where f_i are body forces and t_i are surface tractions.
Substitution of (18.35) into (18.36) gives

$$\int_V \delta\dot{\varepsilon}_{ij}(C^{ep}_{ijkl}\dot{\varepsilon}_{kl})\,dV + \int_V \zeta_{ij}\dot{T}\delta\dot{\varepsilon}_{ij}\,dV = \int_V \dot{f}_i\delta\dot{u}_i\,dV + \int_S \dot{t}_i\delta\dot{u}_i\,dS. \tag{18.37}$$

An application of the Galerkin method to Eq. (18.37) with finite-element discretization of ε and u leads to the equation

$$\mathbf{K}_u\dot{u} + \mathbf{M}_T\dot{T} - L = 0, \tag{18.38}$$

where

$$\mathbf{K}_u = \int_V \mathbf{B}C^{ep}\mathbf{B}\,dV \text{ is the stiffness matrix}, \tag{18.39}$$

$$\mathbf{M}_T = \int_V \mathbf{B}^T\zeta\mathbf{H}\,dV \text{ is the thermomechanical coupling matrix, and} \tag{18.40}$$

$$L = \int_V \mathbf{N}\dot{f}\,dV + \int_S \mathbf{N}\dot{t}\,dS \text{ is the mechanical load vector.} \tag{18.41}$$

18.4.2 Thermo-Rigid Plastic and Thermo-Rigid Visco-Plastic Models

Coupled thermo-rigid plastic and thermo-rigid visco-plastic models are of major practical significance. Descriptions of thermomechanical phenomena in metal forming are usually based upon such models (Służalec 1990b). In these models the elastic effects are neglected and the material law is assumed as follows

$$s_{ij} = 2K \frac{1}{(\sqrt{3}\dot{\bar{\varepsilon}})^{1-m}} \dot{\varepsilon}_{ij} \tag{18.42}$$

where K is the consistency.

In the case $m = 1$, one obtains the stress–strain relation for a rigid visco-plastic body; the case $m = 0$ (and $\dot{\bar{\varepsilon}} \neq 0$) refers to the rigid plastic model. In these models an isotropic hardening law is introduced through a relationship between the consistency K and the equivalent strain $\bar{\varepsilon}$. The most frequently used relations are

$$K(\bar{\varepsilon}) = K_0 \bar{\varepsilon}^n, \tag{18.43}$$

$$K(\bar{\varepsilon}) = K_0 + K_1 \bar{\varepsilon}^n, \tag{18.44}$$

$$K(\bar{\varepsilon}) = K_0(1 + a\bar{\varepsilon})^n, \tag{18.45}$$

where K_0, K_1, a, n are temperature-dependent parameters.

The boundary conditions in metal forming processes have to be modified by introducing the shear stress $\dot{t}$ between tool and specimen through the relation

$$\dot{t} = -\mu K \Delta \dot{u} / |\Delta \dot{u}|^{1-q} \tag{18.46}$$

where μ is the friction coefficient, q is the friction rate sensitivity index, $\Delta \dot{u}$ is the difference of displacement rates in the specimen and the tool.

The finite-element equations are built by considering the universal equilibrium condition for the solid subject to mechanical loads

$$\int_V \sigma_{ij} \delta \dot{\varepsilon}_{ij} \, dV = \int_V f_i \delta \dot{u}_i \, dV + \int_S t_i \delta \dot{u}_i \, dS. \tag{18.47}$$

Introduce the operator $\mathbf{D}$ such that

$$\mathbf{s} = \mathbf{D}\boldsymbol{\sigma}, \quad \boldsymbol{\sigma} = \begin{bmatrix} \sigma_{11} \\ \sigma_{12} \\ \vdots \end{bmatrix}, \quad \text{or} \quad [s_{11} \ s_{12} \ \ldots \ ]^T = \mathbf{s} \tag{18.48}$$

where σ_{ij}, s_{ij} are the components of the stress and deviatoric stress, respectively. Hence and by (18.42)

$$\boldsymbol{\sigma} = \frac{2K}{(\sqrt{3}\dot{\bar{\varepsilon}})^{1-m}} \mathbf{D}^{-1} \dot{\boldsymbol{\varepsilon}}. \tag{18.49}$$

Substitution of the above into Eq. (18.47) gives

$$\mathbf{K}_u \dot{\mathbf{u}} - \mathbf{L} = 0 \tag{18.50}$$

where

$$\mathbf{K}_u = \int_V \mathbf{B}^T \mathbf{D}^{-1} \frac{2K}{(\sqrt{3}\dot{\bar{\varepsilon}})^{1-m}} \mathbf{B} \, dV \text{ is the stiffness matrix and} \qquad (18.51)$$

$$\boldsymbol{L} = \int_V \mathbf{N} \dot{f} dV + \int_S \mathbf{N} \dot{t} \, dS \text{ is the mechanical load vector.} \qquad (18.52)$$

The values of $\dot{u}$ can be updated as follows

$$\dot{u}^{n+1} = \dot{u}^n + \theta \Delta \dot{u} \qquad (18.53)$$

where $0 \leq \theta \leq 1$ is chosen such as to ensure the convergence of the iterative sequence. Convergence criteria can be divided into two parts. The first one analyses the norm

$$\frac{\|\Delta \dot{u}\|}{\|\dot{u}\|} \qquad (18.54)$$

where $\|\cdot\|$ denotes the Euclidean norm. The second one considers the norm of residual equations (18.50). The first criterion is usually used at the beginning of the iteration process. The second one is applied for the last iterations.

18.4.3 Remarks on Other Models

Many different models of constitutive laws have been proposed in order to describe the inelastic behaviour of solid bodies in coupled thermomechanical processes. In this book we cannot give a comprehensive survey of all existing theories; only a selection can be described. The coupled thermomechanical problem can be analysed for example with a model based on multiplicative decomposition of the deformation gradient (Wriggers et al. 1990) which was briefly described in Section 17.2. The other models which should be mentioned here are: the thermo-visco-plasticity model of Cernocky and Krempl (1980a, b), which is related primarily to creep and relaxation processes; Hart's (1976, 1978, 1982) theory of thermo-visco-plasticity, which is intended to cover the whole field of non-elastic deformations, i.e. visco-plastic processes as well as thermally activated creep and relaxation processes; and Raniecki's (1983) theory of thermo-plasticity, which is restricted to non-isothermal elasto-plastic deformations.

18.5 Coupled Thermomechanical Algorithm

In the case when complete coupling is considered, the two unknown quantities, the deformation rate vector $\dot{u}$ and temperature rate vector $\dot{T}$, are coupled and should be determined simultaneously. In coupled thermo-elasto-plastic problems we have to consider together Eqs. (18.27) and (18.38). Thus

$$\begin{bmatrix} \mathbf{K}_u & \mathbf{M}_T \\ \mathbf{M}_u & \mathbf{C} \end{bmatrix} \begin{Bmatrix} \dot{u} \\ \dot{T} \end{Bmatrix} = \begin{Bmatrix} L \\ P \end{Bmatrix} \qquad (18.55)$$

where

$$P = D + Q - K_T T.$$

We introduce the non-symmetric matrix K_{TM}

$$K_{TM} = \begin{bmatrix} K_u & M_T \\ M_u & C \end{bmatrix}. \tag{18.56}$$

The solution strategy often used neglects the coupling terms M_T and M_u and hence yields the uncoupled operator

$$K_{TM}^s = \begin{bmatrix} K_u & 0 \\ 0 & C \end{bmatrix}. \tag{18.57}$$

The consistent linearization which leads to the mechanical part K_u, is performed by holding the temperature fixed. Analogously the thermal part in (18.54) is derived by fixing the deformation constant. There exist some alternative solution strategies depending on the kind of thermomechanical process. The most commonly used algorithm solves in the first loop the mechanical equations using the temperatures known from the last converged solution. Then a loop over the thermal and mechanical equations is performed until convergence. The thermal equations are solved first based on the uncoupled operator. In the next step the mechanical equations are solved with the already updated thermal variables (Służalec 1990c).

18.6 Examples

As an example, an analysis of ring compression is presented (Służalec 1988d). Let us consider a scheme of ring compression given in Fig. 18.1. A finite-element mesh consisting of four nodal isoparametric elements within the quarter of the ring is presented in Fig. 18.2. Beginning with a temperature of 900 °C a steel ring forged under various friction conditions is modelled to investigate the influence of friction conditions on the accuracy of the shape. The yield stress of the steel is given in Fig. 18.3. In the investigations friction coefficients of 1 and 0.3 have been chosen. The initial temperature of the die varies from 15 °C to 900 °C. The assumed velocity of flow is 20 mm s^{-1}. The thermal parameters of materials used for calculations are given in Table 18.1. In Fig. 18.4 the material flows after 70% height reduction are plotted for different forming conditions. Fig. 18.5 shows temperature fields in the ring and the die for the same conditions as in Fig. 18.4.

In the next example we consider a coupled thermomechanical process during compression test (Bruhns and Służalec 1989). A rigid-plastic material model is assumed. The dimensions of the specimen are shown in Fig. 18.6 and the finite-element mesh used, in Fig. 18.7. For computations we assumed

$$\sigma = a + c\bar{\varepsilon}^n \tag{18.58}$$

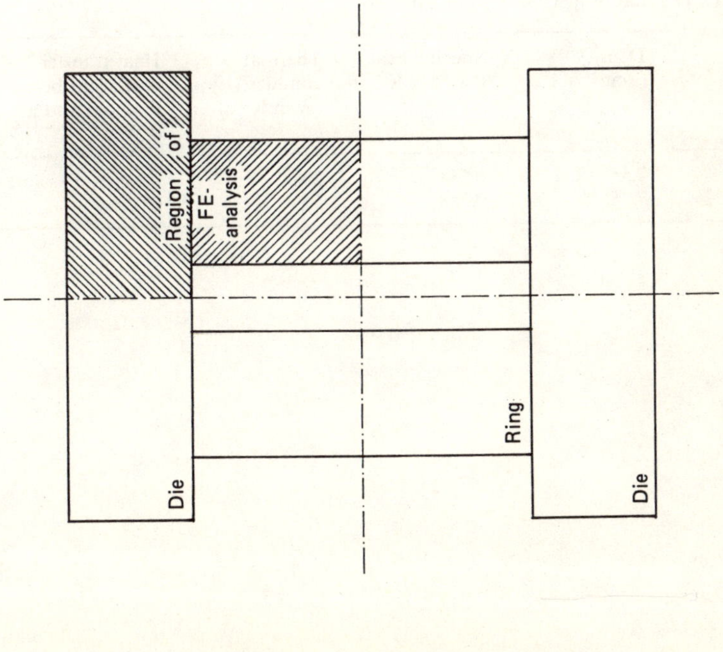

EQUATIONS OF COUPLED THERMO-PLASTICITY   171

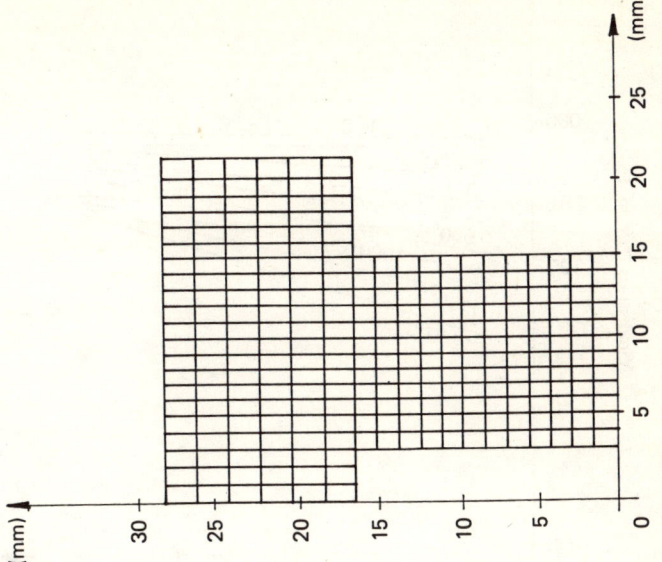

Figure 18.2. Finite-element mesh for ring compression.

Figure 18.1. Modelling of ring compression.

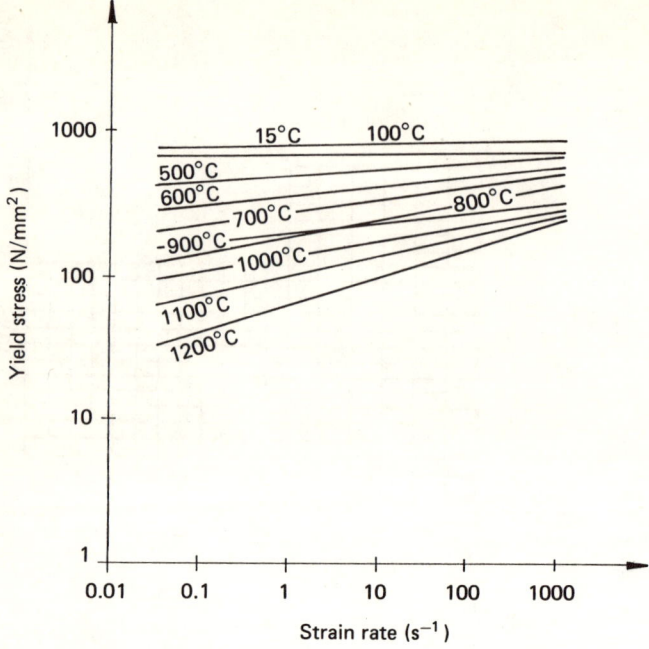

Figure 18.3. Yield stress for the assumed material.

Table 18.1. Thermal properties of material used

Material	Density (g cm^{-3})	Specific heat (kJ kg^{-1} K^{-1})	Thermal conductivity (W m K^{-1})	Heat transfer coefficient between material and air (kW m^{-2} K^{-1})
Ring	7.55	0.63	29	0.035
Die	7.75	0.588	54	0.027

EQUATIONS OF COUPLED THERMO-PLASTICITY 173

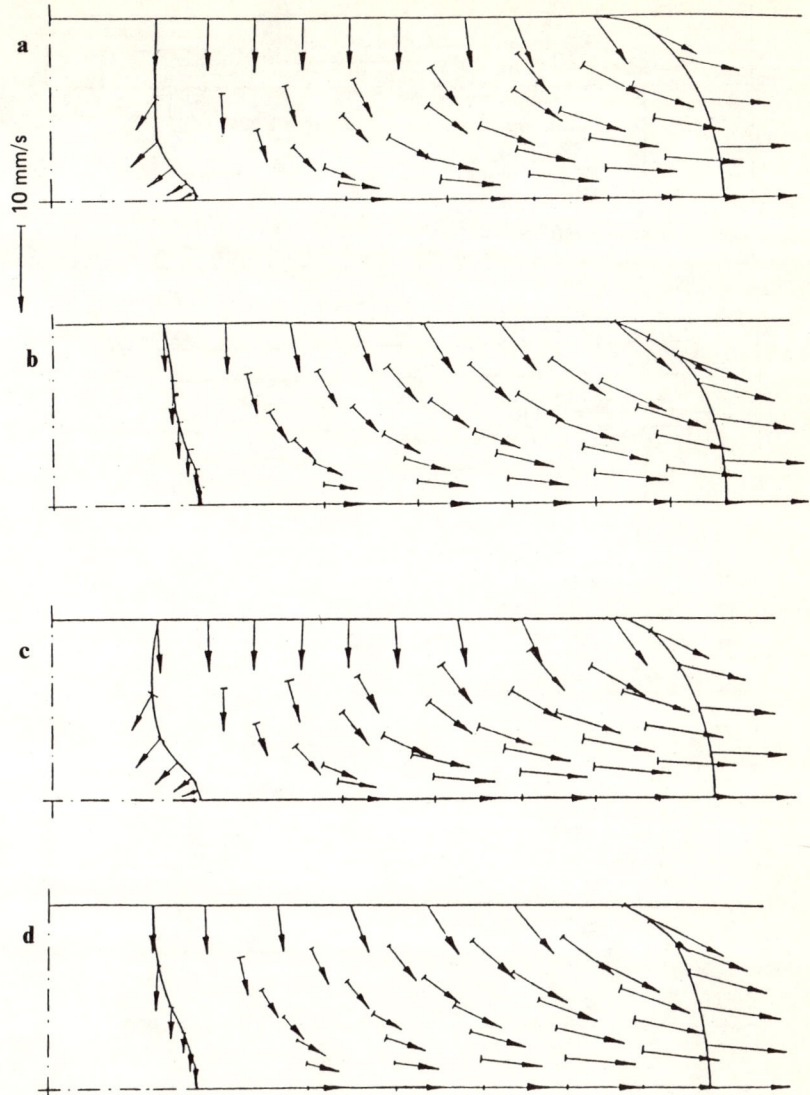

Figure 18.4. Metal flow in the chosen nodes of the ring for 70% reduction and conditions given below:

Figure	a	b	c	d
Die temperature (°C)	15	15	900	900
Friction coefficient	1	0.3	1	0.3

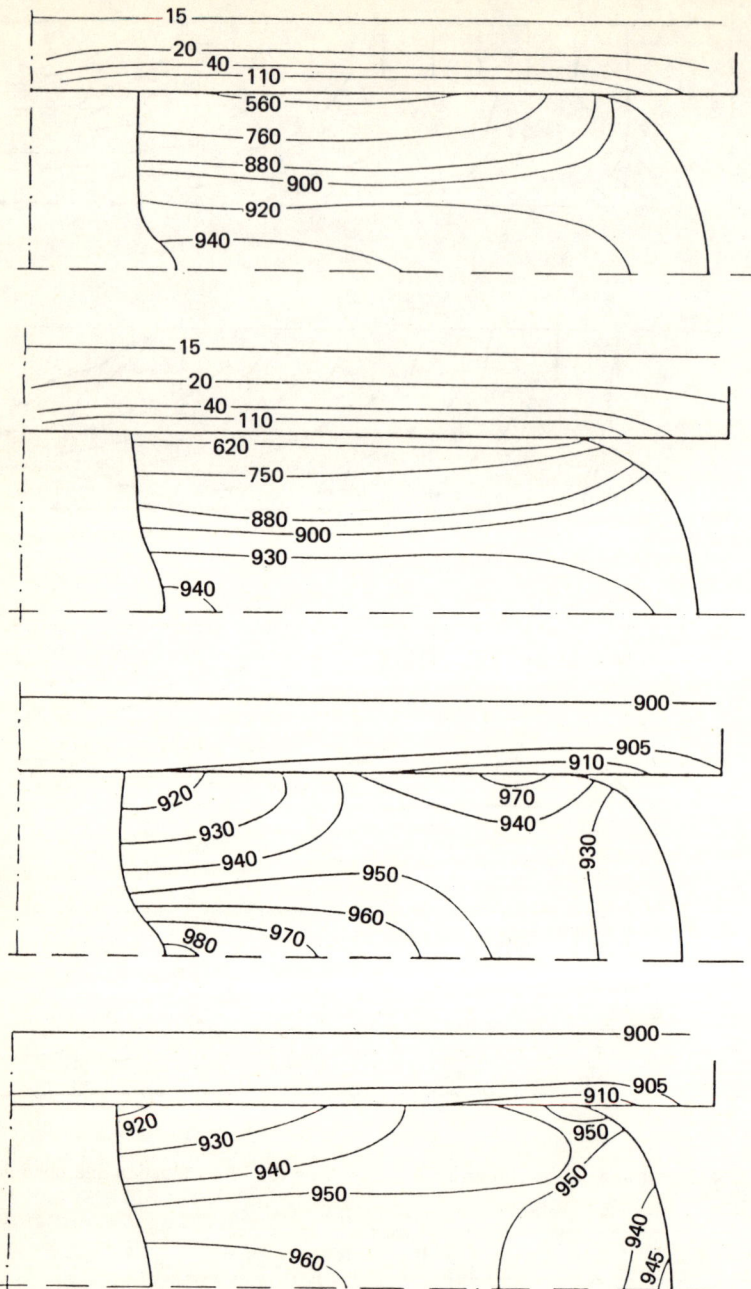

Figure 18.5. Temperature fields in the ring and the die (70% reduction) for the same conditions as in Fig. 18.4.

EQUATIONS OF COUPLED THERMO-PLASTICITY 175

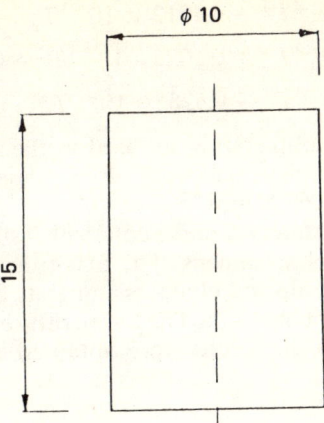

Figure 18.6. Dimensions of the specimen.

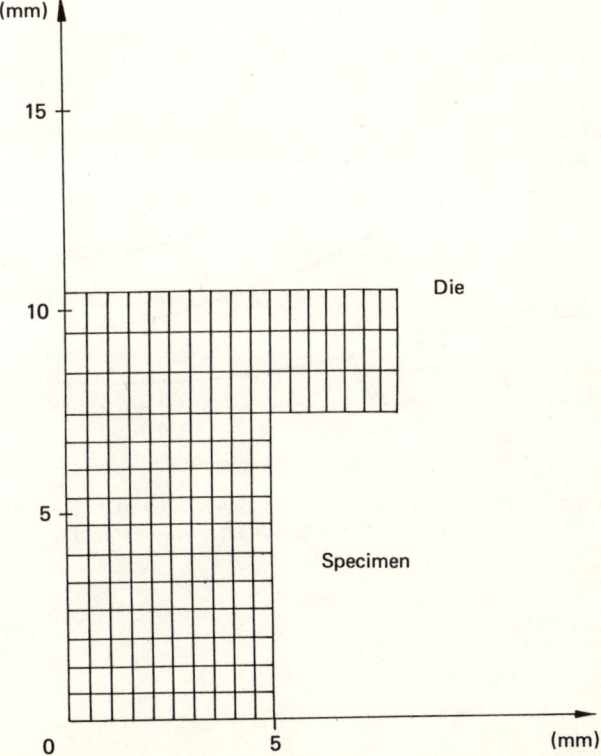

Figure 18.7. Finite-element mesh used.

where $a = -7.2588 \times 10^{-8}T^3 + 4.4912 \times 10^{-4}T^2 - 0.938T - 64.6$,

$n = -0.1743 \times 10^{-9}T^3 + 0.1026 \times 10^{-5}T^2 - 0.1917 \times 10^{-2}T + 1.3$,

$c = -0.8682 \times 10^{-8}T^3 + 0.5285 \times 10^{-4}T^2 - 0.107T + 72.28$.

The following thermal parameters were used in the simulation: environmental temperature 20 °C, initial temperature 20 °C, specimen heat conductivity 30.28 N s^{-1} K^{-1}, $\rho C = 3.69$ N mm^{-2}.

Fig. 18.8 presents experimental and numerical results regarding temperature changes on the surface of specimens. The experiments were carried out on a universal testing machine during compression tests and assuming velocities of deformation 10, 50 and 150 mm s^{-1}. The temperature was monitored by using a thermovision camera. All the results were obtained at room temperature.

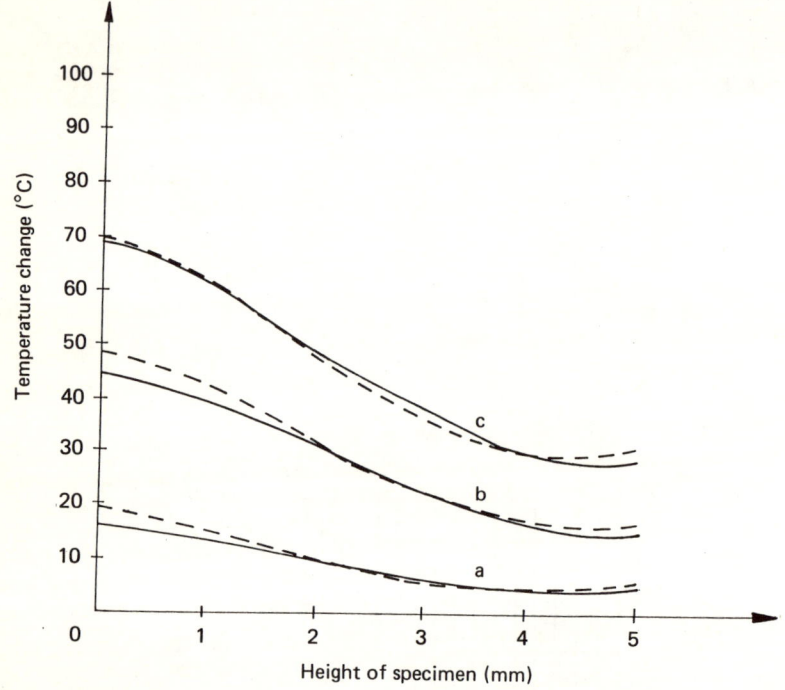

Figure 18.8a–c. Temperature changes in the specimen after 1/3 reduction: **a** 10 mm s^{-1}; **b** 50 mm s^{-1}; **c** 150 mm s^{-1}. *Continuous line*, experiment; *broken line*, numerical solution.

References and Further Reading

Almansi E (1911) Sulle deformazione finite del solidi elastici istropii: I. *Rendi conti Accad Naz Lineei* 20(5): 704–714

Andrade EN da C (1910) On the viscous flow of metals and allied phenomena. *Proc R Soc Lond Ser A* 84:567A

Bailey RW (1929) Creep of steel under simple and compound stresses and the use of high initial temperature in steam power plant. Trans Tokyo Sectional Meeting of the World Power Conf, p. 1089

Bathe KJ (1982) Finite element procedures in engineering. Prentice-Hall, Englewood Cliffs

Bathe KJ, Slawkowic R, Kojic M (1986) On large strain elastic–plastic and creep analysis. In: Finite element methods for nonlinear problems. Springer, Berlin Heidelberg New York, pp 175–190

Bauschinger J (1881) Über die Veränderung der Elastizitätsgrenze und des Elastizitätsmodulus verschiedener Metalle. *Civilingenieur*, pp 289–348

Bruhns OT, Służalec A (1989) Thermal effects in thermo-plastic metal with internal variables. *Computers & Structures* 33(6):1459–1464

Cauchy AL (1827) De la pression en tension dans un corps solide. *Ex de Math* 2:42–56

Cernocky EP, Krempl E (1980a) A theory of viscoplasticity based on infinitesimal total strain. *Acta Mech* 36:263–289

Cernocky EP, Krempl E (1980b) A theory of thermoviscoplasticity based on infinitesimal total strain. *Int J Solids Struct* 16:723–741

Clark DS, Duwez PE (1950) The influence of strain rate on some tensile properties of steel. *Proc Am Soc Testing Mat* 50:560–675

Comini G, Del Guidice S, Lewis RW, Zienkiewicz OC (1974) Finite element solution of non-linear heat conduction equations with special reference to phase change. *Int J Numer Meth Eng* 8:613–624

Cowper GR, Symonds PS (1957) Strain hardening and strain rate effects in the impact loading of a cantilever beam. *Technical report 28*, Brown University

Crank J. (1981) How to deal with moving boundaries in thermal problems. In: Lewis RW, Morgan K, Zienkiewics OC (eds) *Numerical methods in heat transfer*. Wiley, Chichester

Dafalias YF (1985) The plastic spin. *J Appl Mech* 52:865–871

D'Alembert JL (1752) Essai d'une nouvelle théorie de la résistance des fluides. Paris

Dashner PA (1986) Invariance considerations in large strain elasto-plasticity. *J Appl Mech* 53:55–60

Davenport CC (1938) Correlation of creep and relaxation properties of copper. *J Appl Mech* 5:2

Davis EA (1937) Plastic behaviour of metals in the strain-hardening range. II. *J Appl Phys* 8:213–217

Davis EA (1943) Increase of stress with permanent strain and stress–strain relations in the plastic state for copper under combined stress. *J Appl Mech* 10:A187–A196

Davis EA (1961) The Bailey flow rule and associated yield surface. Trans *ASME Ser E* 28(2):310

Delang L, Krishan RV, Tas H, Warlimont H (1974) Thermoelasticity, pseudoplasticity and the memory effects associated with martensitic transformation. *J Mat Sci* 9:1521–1535, 1536–1544, 1545–1555

Dillon OW Jr (1962a) A nonlinear thermoelasticity theory. *J Mech Phys Solids* 10:123–131

Dillon OW Jr (1962b) An experimental study of the heat generated during torsional oscillations. *J Mech Phys Solids* 10:235–244

Dillon OW Jr (1963) Coupled thermoplasticity. *Nucl Eng Des* 24:1–55

Donea J, Giuliani S, Laval H, Quartapelle L (1981) Solution of the unsteady Navier–Stokes equations by a finite element projection method. In: Taylor C, Morgan K (eds) *Recent advances in numerical methods in fluids*, vol. 2: Computational techniques in transient and turbulent flow. Pineridge Press, Swansea

Dorn JE (1955) Some fundamental experiments on high temperature creep. *J Mech Phys Solids* 3(2):85

Dorn JE, Tietz TE (1949) Creep and stress rupture investigations on some aluminium alloy sheet metals. *Proc ASTM* 49

Drucker DC (1951) A more fundamental approach to plastic stress–strain relations. *Proc First US Nat Congr Appl Mech*, Chicago, pp 487–491

Drucker DC (1959) A definition of stable inelastic material. *J Appl Mech* 26(1):101–106

Drucker DC, Hopkins HG (1955) Combined concentrated and distributed load on ideally plastic circular plates. *Proc Second US Nat Congr Appl Mech*, Ann Arbor, Michigan 1954, pp 517–520

Dupont T, et al. (1974) Three-level Galerkin methods for parabolic equations. *SIAM J Numer Anal* 11:392–410

Ericksen JL (1960) Tensor fields. In: Flügge S (ed) *Encyclopedia of physics*, vol III/1. Springer, Berlin Heidelberg New York

Farren WS, Taylor GI (1925) The heat developed during plastic extension of metals. *Proc R Soc Lond* 107:442

Findley WN (1944) Creep characteristics of plastic (symposium on plastics). *Proc ASTM* 44:118

Findley WN, Khosla G (1956) An equation for tension creep of three unfilled thermoplastics. *Soc Plastics Eng J* 12(20)

Findley WN, Peterson DB (1958) Prediction of long-time creep with ten-year creep data on four plastic laminates. *Proc ASTM* 58: 841–853

Findley WN, Adams CH, Worley WJ (1948) The effect of temperature on the creep of two-laminated plastics as interpreted by the hyperbolic-sine law and activation energy theory. *Proc ASTM* 48:1217

Findley WN, Lai JS, Onaran K (1976) Creep and relaxation of nonlinear viscoplastic materials. North-Holland, Amsterdam

Garofalo F (1965) Fundamentals of creep and creep rupture in metals. Macmillan, New York

Gartling DK (1980) Finite element analysis of convective heat transfer problems with change of phase. In: Morgan K, Taylor C, Brebbia CA (eds) *Computer methods in fluids*. Pentech, London

Graham A, Walles KFA (1955) Relation between long and short time properties of a commercial alloy. *J Iron Steel Inst* 179

Grant NJ, Bucklin AG (1965) On the extrapolation of short time stress rupture data. In: Grant NJ, Mullendore AW (eds) MIT Press

Green AE, Naghdi PM (1965) A general theory of an elastic-plastic continuum. *Arch Rat Mech Anal* 18(261):81

Green AE, Naghdi PM (1971) Some remarks on elastic-plastic deformation at finite strain. *Int J Eng Sci* 9:1219

Green G (1841) On the propagation of light in crystallized media. *Trans Cambridge Phil Soc* 7:121–140

Guest JJ (1900) On the strength of ductile materials under combined stress. *Phil Mag* 50:69–132
Hart EW (1976) Constitutive relations for the nonelastic deformation of metals. *Trans ASME J Eng Mat Tech* 98:193–202
Hart EW (1978) Constitutive relations for the nonelastic deformations. *Nucl Eng Des* 46:179–185.
Hart EW (1982) The effects of material rotations in tension-torsion testing. *Int. J Solids Struct* 18:1031–1042
Hauser FE, Simmons IA, Dorn IE (1960) Strain rate effects in plastic wave propagation, Technical report no. 3, University of California
Hencky H. (1924) Zur Theorie plastischer Deformationen und der hierdurch in Material hervorgerufen Nach-Spannungen. *ZAMM* 4:324–334
Hill R (1978) Aspects of invariance in solid mechanics. *Adv Appl Mech* 18:1–75
Hoff NJ 1953 The necking and rupture of rods subjected to constant tensile loads. *J Appl Mech* 20
Hohenemser K, Prager W (1932) Beitrag zur Mechanik des bildsamen Verhaltens von Flussstahl. *ZAMM* 12(1):1–14
Hsu TR (1986) The finite element method in thermomechanics. Allen and Unwin, Boston
Huber MT (1956) Właściwa praca odkształcenia jako miara wytężnia materiału. *Czas Techn* 1904(22):34–40, 49–50, 61–62, 80–81, *Lwów, Pisma t. II*, Warsaw
Hughes TJR (1980) Finite rotation effects in numerical integration of rate constitutive equations arising in large-deformation analysis. *Int J Num Meth Eng* 15:1413–1418
Hughes TJR (1984) Numerical implementation of constitutive models: rate-independent deviatoric plasticity. In: Theoretical foundations for large scale computations of nonlinear material behaviour. Martinus Nijhoff, Dodrecht, pp 29–57
Hughes TJR, Winget J (1980) Finite rotation effects in numerical integration of rate constitutive equations arising in large-deformation analysis. *Int J Num Meth Eng* 15:1413–1418
Iljuszyn AA (1943) Some problems of plastic deformations. *Prikl Mat Mech* 7(4):245–272
Iljuszyn AA (1946) On theory of small elasto-plastic deformations. *Prikl Mat Mech* 10(3):347–356
Iljuszyn AA (1948) Plasticity. Gos Izd Tech Teor Lit Moscow Leningrad
Ivey HJ (1961) Plastic stress–strain relations and yield surfaces for aluminium alloys. *J Mech Eng Sci* 3(1):15–31
Johnson AE, Henderson J, Khan B (1963) Multiaxial creep strain–complex stress–time relations for metallic alloys with some applications to structures. *Proc int conf on creep*, ASME (ASTM). *J Mech Eng* 26
Kachanov LM (1958) On rupture time in creep conditions. *Izw AH USSR, OTN* 8:26–31
Kachanov LM (1960a) On rupture time in creep conditions. *Izw AH USSR, OTN* 5:89–92
Kachanov LM (1960b) Creep theory. Fizmagiz, Moscow
Kachanov LM (1961) Creep rupture time: problems of continuum mechanics (in honour of the seventienth birthday of N.I. Muskeliszwili). *Soc Ind Appl Math*, Philadelphia, pp 202–221
Kauzman W (1941) Flow of solid metals from the standpoint of the chemical rate theory. *Trans AIME* 1943
Kirchhoff G (1852) Über die Gleichungen des Gleichgewichts eines elastischen Körpers bei nicht unendlich kleinen Verschiebungen seiner Teille. *Sitz Acad Wiss*, Vienna 9:762–773
Kleiber M, Służalec A (1983a) Numerical analysis of heat flow in flash welding. *Arch Mech* 35(5–6):687–699
Kleiber M, Służalec A (1983b) A numerical algorithm for nonlinear diffusion problem and its application to carbon diffusion analysis during steel carburizing. *J Tech Phys* 24(4):455–466
Kleiber M, Służalec A (1984) Finite element analysis of heat flow in friction welding. *Eng Trans* 32(1):107–113
Larson FR, Miller J (1952) A time–temperature relationship for rupture and creep stresses. *Trans ASSME* 74(5)
Leaderman H (1943) Elastic and creep properties of filamentous materials. Textile Foundation, Washington DC
Lee EH (1969) Elastic plastic deformations at finite strains. *J Appl Mech* 36:1–6

Lee EH, Mallet RL, Wertheimer TB (1983) Stress analysis for anisotropic hardening in finite-deformation plasticity. *J. Appl Mech* 50:554–560

Lehmann Th (1960) Einige Betrachtungen zur Beschreibung von Vorgängen in der Klassichen Kontinuums-mechanik. *Ing Arch* 29:316–330

Lehmann Th (1962) Einige ergänzende Betrachtungen zur Beschreibung von Vorgängen in der Klassichen Kontinuums-mechanik. *Ing Arch* 31:371–384

Lehmann Th (1973) On large elastic–plastic deformations. In: Foundations of plasticity. Nordhoff, Leyden, pp 571–585

Lehmann Th (1977) On the theory of large, non-isothermic, elastic–plastic and elastic–viscoplastic deformation. *Arch Mech* 29(3):393–409

Lehmann Th (1979) Coupling phenomena in thermoplasticity. In: Trans. 5th Int Conf Structure Mech Reactor Technol, Berlin, paper L1/1

Lehmann Th (1980) Coupling phenomena in thermoplasticity. *Nucl Eng Des* 57:323–32

Lehmann Th (1982a) Some remarks on the decomposition of deformations and mechanical work. *Int J Eng Sci* 20:281–288

Lehmann Th (1982b) Some theoretical considerations and experimental results concerning elastic–plastic stress–strain relations. *Ing Arch* 52:391–403

Lehmann Th (1983a) General frame for the definition of constitutive laws for large non-isothermic elastic–plastic and elastic–viscoplastic deformations. In: The constitutive law in thermoplasticity. *CISM Courses and lectures*. Springer, Berlin Heidelberg New York

Lehmann Th (1983b) Einige Aspekte der Thermoplastizität. *ZAMM* 63(3/13)

Lehmann Th (1984) Some considerations on the constitutive law in thermoplasticity. *Mechanika Teoretyczna i Stosowana* 1–2(22):3–20

Levy M (1870) Memoire sur les equations generales. *C R Acad Sci* 70:1323–1325

Lifszic IM (1963) On the theory of diffusive mechanisms of creep in polycrystalline bodies (in Russian). *Z Eksp Techn Fiz* 44(4)

Loret B (1983) On the effects of plastic rotation in the finite deformation of anisotropic materials. *Mech Mat* 2:287–304

Ludwik P (1909) Elemente der technologischen Mechanik. Springer Berlin Heidelberg New York

Maiden CI, Campbell ID (1958) The static and dynamic strength of a carbon steel at low temperature. *Phil Mag* 3:872–885

Malinin NN (1975) The theory of plasticity and creep (in Russian). Maszinostrojenije, Moscow

Malinin NN, Rżysko J (1981) Mechanics of materials (in Polish). PWN, Warsaw

Malvern LE (1951a) Plastic wave propagation in a bar of material exhibiting a strain rate effect. *Q Appl Math* 8:405–411

Malvern LE (1951b) The propagation of longitudinal waves of plastic deformation in a bar of material exhibiting a strain rate effect. *J Appl Mech* 18(3):203–208

McVetty PG (1934) Working stress for high temperature service. *Mech Eng* 56

McVetty PG (1943) Creep of metals at elevated temperatures – the hyperbolic sine relation between stress and strain rate. *Trans ASME* 65

Melan E. (1938) Zur Plastizität der räumlichen Kontinuums. *Ing. Archiv.* 9:116–126

Meyer GH (1978) The numerical solution of multidimensional Stefan problems – a survey. In: Wilson DG, Solomon AD, Boggs PT (eds) Moving boundary problems. Academic Press, New York

Miller I (1983) Thermodynamic theories, thermoplasticity and special cases of thermoplasticity. In: CISM courses and lectures. Springer, Berlin Heidelberg New York

Mises R (1913) Mechanik der festen Körper im plastisch deformablen Zustand. *Göttingen Nachrichten, Math Phys* 4(1):582–592

Mises R (1928) Mechanik der plastischen Formänderung von Kristallen. *ZAMM* 8(3):161–185

Morgan K (1980) A numerical analysis of freezing and melting with convection. *Comp Meth Appl Mech Eng* 15:1413–1418

Morgan K, Lewis RW, Zienkiewicz OC (1978) An improved algorithm for heat conduction problems with phase change. *Int J Numer Meth Eng* 13:1191–1195

Morgan K, Lewis RW, Thomas HR (1980) Finite element modelling of drying stresses in timber and cooling stresses in cast metal. In: O'Carroll MJ, et al. (eds) Modelling and simulation in practice, vol 2. Emjoc, Northallerton

Mroz Z, Raniecki B (1976) On the uniqueness problem in coupled thermo-plasticity. *Int J Eng Sci* 14:211–221

Nadai A (1923) Der Beginn des Fliessvorganges in einen tordierten Stab. *ZAMM* 3(6):442–444

Nadai A (1931) Plasticity. McGraw-Hill, New York

Nadai A (1937a) Plastic behaviour of metals in the strain-hardening range: I. *J Appl Phys* 8:205–213

Nadai A (1937b) On the creep of solids at elevated temperatures. *J Appl Phys* 8:418–425

Nadai A (1938) The influence of time upon creep; the hyperbolic sine creep law. *S. Timoshenko anniversary volume*. Macmillan, New York, p 155

Nadai A (1950) Theory of flow and fracture of solids. McGraw-Hill, New York

Nagtegaal JC, Veldpaus FE (1984) On the implementation of finite strain plasticity equation in a numerical model. In: Numerical analysis of forming processes. Wiley, Chichester, pp 351–357

Nemat-Nasser S (1982) On finite strain elasto-plasticity. *Int J Solids Struct* 18:857–872

Nemat-Nasser S (1983) On finite plastic flow of crystalline solids and geometrials. *J Appl Mech* 50:1114–1126

Nied HA, Batterman SC (1972) On the thermal feedback reduction of latent energy in the heat conduction equation. *Mat Sci Eng* 9:243–245

Norton FH (1929) The creep of steel at high temperatures. McGraw-Hill, New York

Oden JT (1972) Finite elements in nonlinear continua. McGraw-Hill, New York

Oding IA (1959) Creep theory of metals (in Russian). Moscow

Odqvist FKG (1933) Die Verfestigung von flusseisenähnlichen Körpern. *ZAMM* 13(5):360–363

Odqvist FKG (1961) On theories of creep rupture. Institutionen för Höllfasthetslära, Kungl. Tekniska Högskolan, No. 136

Odqvist FKG (1966) Mathematical theory of creep and creep rupture. Clarendon Press, Oxford

Odqvist FKG, Hult J (1961) Some aspects of creep rupture. *Arkiv för Fysik* 19:379–382

Odqvist FKG, Hult J (1962) Kriechfestigkeit metallischer Werkstoffe. Springer, Berlin Heidelberg New York

Oldroyd JG (1950) On the formulation of the rheological equations of state. *Proc R Soc Lond, Ser A* 200:523–541

O'Neil K, Lynch DR (1981) A finite element solution of freezing problems using a continuously deforming coordinate system. In: Lewis RW, Morgan K, Zienkiewicz OC (eds) *Numerical methods in heat transfer*. Wiley, Chichester

Penny RK, Marriott DL (1971) Design for creep. McGraw-Hill, Maidenhead, UK

Perzyna P (1966) Theory of viscoplasticity (in Polish). PWN Warsaw.

Philips A (1956) Introduction to plasticity. Ronald Press, New York

Philips P (1950) The slow stretch in india rubber, glass and metal wires when subjected to a constant pull. *Phil Mag* 9:513–531

Piola G (1833) La meccanica de'corpi naturalmente estesi trattata col calcolo delle variazioni. *Opusc Mat Fis Di Diversi Autori Milano Guisti* 1:201–236

Podgornyj AN (1984) Creep of machine tool elements (in Russian). USSR Academy of Science, Kiev

Prager W (1938) On isotropic materials with continuous transition from elastic to plastic state. *Proc Fifth Int Congr Appl Mech* Cambridge, Mass, pp 234–237

Prager W (1948) Discontinuous solutions in the theory of plasticity. Courrant anniversary volume. Interscience, New York, pp 289–299

Prager W (1955a) The general theory of limit design. *Proc Eighth Int Congr Appl Mech* 1952, Istanbul 2:65–72

Prager W (1955b) The theory of plasticity: a survey of recent achievements. *Proc Inst Mech Eng* London

Prager W (1957) Total creep under varying loads. *J Aero Sci* 24(2):154
Prager W, Hodge PG (1951) Theory of perfectly plastic solids. Wiley, New York
Prandtl L (1923) Anwendungsbeispiele zu einem Henckyschen Satz über das plastische Gleichgewicht. *ZAMM* 3(6):401–406
Prandtl L (1924) Spannungsverteilung in plastischen Körpern. *Proc First Int Congr Appl Mech*, Delft, pp 43–54
Prandtl L (1928) Ein Gedenkenmodell zur kinetishen Theorie der festen Körper. *ZAMM* 8:85–106
Rabotnov YuN (1948) Strength of elements in creep conditions. *Izd. AN CCCP*, Otd, Tech N
Rabotnov YuN (1966) Creep of machine structural elements (in Russian). Nauka, Moscow
Rabotnov YuN (1968) Creep rupture. *Proc 12th Int Congr Appl Mech*, Stanford, pp 342–349
Ramberg G, Osgood W (1943) Description of stress–strain curves by three parameters. *Nat Adv Comm for Aeronautics*, Tech Note No. 902
Raniecki B (1983) Thermodynamic aspects of cyclic and monotone plasticity. In: The constitutive law in thermoplasticity. *CISM courses and lectures*. Springer, Berlin Heidelberg New York
Raniecki B, Sawczuk A (1975) Thermal effects in plasticity: I. Coupled theory. *Z Angew Math Mech* 55:333–341
Reed KW, Atluri SN (1985) Constitutive modeling and computational implementation for finite strain plasticity. *Int J Plasticity* 1:63–87
Reiner M (1945) A mathematical theory of dilatancy. *Am J Math* 67:350–362
Reuss A (1933a) Brücksichtigung der elastischen Formänderungen in der Plastizitätstheorie. *ZAMM* 10(3):266–274
Reuss A (1933b) Vereinfachte Berechnung der plastischen Formänderungsgeschwindigkeiten bei Voraussetzung der Schubspannungsfliessbedingung. *ZAMM* 13(5):356–360
Rice JR (1975) Continuum mechanics thermodynamics of plasticity in relation to microscale deformation mechanism. In: *Constitutive equations in plasticity*. MIT Press. Cambridge, Mass, pp 23–75
Rolph III WD, Bathe KJ (1984) On a large strain finite element formulation for elastoplastic analysis. In: Constitutive Equations: macro and computational aspects. *ASME winter annual meeting*, pp 131–147
Rubinstein R, Atluri SN (1983) Objectivity of incremental constitutive relations over finite time steps in computational finite deformation analysis. *Comp Meth Appl Mech Eng* 36:277–290
Sdobyriev WR (1959) Strength criteria for some heat resistant alloys under complex stress state. *Izd AN USSR OTN Mech Machinostrojenije*, pp 93–99
Seth BR (1962) Generalized strain measure with applications to physical problems. *Proc IUIAM Symp. on Second-order effects*, Haifa. Pergamon Press, Oxford
Simo JC, Taylor RL (1986) A return mapping algorithm for plane stress elasto-plasticity. *Int J Num Meth Eng* 22:649
Skrzypek J (1986) Plasticity and creep. PWN, Warsaw
Służalec A (1985) Numerical analysis of heat flow under freezing conditions in groundwater system, 91–96. *Acta Geophys Pol* 33(1):83–90
Służalec A (1986a) Heat flow within the eye in cryotherapy. Transactions of the Fifth International Conference on Mechanics in Medicine and Biology, Bologna, Italy II:277–281
Służalec A (1986b) Stresses, deformations and creep-damage of turbine-rotor disc. *Int J Mech Sci* 28(7):443–453
Służalec A (1987a) Thermal effects in laser microwelding. *Computers & Structures* 25(1):29–34
Służalec A (1987b) Thermo-elastic stresses within a rectangular conductor carrying an alternating current. *Comp Meth Appl Mech Eng* 67:253–264
Służalec A (1987c) Finite element model of heat flow in biological tissue undergoing laser irradiation. *J Biomech* 20(10):937–941
Służalec A (1988a) Temperature field within induction heating element. *Int J Eng Sci* 26(3):285–291

Służalec A (1988b) Thermal effects within the eye in cryotherapy. *Computers & Structures* 29(4):661–665
Służalec A (1988c) Flow of metal undergoing laser irradiation. *Num Heat Transfer* 14:253–263
Służalec A (1988d) An analysis of thermal effects of coupled thermoplasticity in metal forming processes. *Comm Appl Num Meth* 4:675–685
Służalec A (1988e) Errors in one-step schemes for solution of the heat flow equation. *Appl Math Modelling* 12:491–494
Służalec A (1989a) Groundwater flow effects in processes of soil freezing *Num Heat Transfer* 15:399–409
Służalec A (1989b) An analysis of thermal phenomena in electromagnetic field during electroslag welding. *Int J Computers Fluids* 17(2):411–418
Służalec A (1989c) Shape optimization of weld surface. *Int J Solids Struct* 25(1): 23–31
Służalec A (1990a) An evaluation of the internal dissipation factor in coupled thermo-plasticity. *Int J Nonlinear Mech* 25(4):395–403
Służalec A (1990b) An application of numerical simulation in the design of metal forging process (in Polish). *Mechanika i Komputer* (in press)
Służalec A (1990c) Thermal effects in friction welding. *Int J Mech Sci* 32(6):467–478
Służalec A (1991) Temperature field in random conditions. *Int J Heat Mass Transfer* 31(1):55–58
Służalec A, Kysiak A (1991) An analysis of weld geometry in creep of welded tube undergoing internal pressure. *Computers & Structures* (in press)
Służalec A, Muskalski K (1985) The finite element method – a new approach to determination of heat flow in the eye shown in the example of cryotherapy: preliminary report. *Jap J Ophthalmol* 29(1):24–30
Soderberg CR (1936) The interpretation of creep test for machine design. *Trans ASME* 58:733–751
Sokołowski WW (1950) Creep theory (in Russian). Gostieizdat, Moscow
St Venant AJCB de (1844) Sur les pressions qui se développent à l'interieur des corps solides lorsque les déplacements de leur points, sans alterer l'élasticité, ne peuvent cependant pas être considérés comme trèspetits. *Bull Soc Philomats* 5:26–28
St Venant AJCB de (1897) Mémoire sur l'equilibre des corps solides, dans les limites de leur élasticité, et sur les conditions de leur résistance, quand les déplacements ne sont pas trés petits. *CR Acad Sci* 24:260–263
Szor BF (1958) Influence of heating on changes of stress state in creep conditions (in Russian). *DAN USSR* 123(5)
Taylor GI, Quinney H (1931) The plastic distortion of metals. *Phil Trans R Soc* Ser A 230:323–362 230:323–362
Treska H (1868) Mémoire sur l'écoulement des corps solides. *Mem Par Div Sav* 18:733
Truesdell C, Toupin R (1960) The classical field theories. *Handbuch der Physik*, Vol III/1. Springer, Berlin Heidelberg New York
Volterra E (1951) On elastic continua with hereditary characteristics. *J Appl Mech* 18
Washizu K (1982) Variational methods in elasticity and plasticity. Pergamon Press, Oxford
Wilkins ML (1964) Calculation of elastic–plastic flow, In: Alder B et al. (eds) Methods of computational Physics, vol 3. Academic Press, New York
Wriggers P, Miehe C, Kleiber M, Simo JC (1990) On the coupled thermo-mechanical treatment of necking problems via finite element method. *Proc Second Int Conf Comput Plasticity*, part I:527–542
Ziegler H (1959) A modification of Prager's hardening rule. *Q Appl Math* 17(1):55–65
Zienkiewicz OC, Cormeau IC (1974) Viscoplasticity, plasticity and creep in elastic solids – a unified numerical approach. *Int J Num Meth Eng* 8:821–845

Subject Index

Acceleration of particle 19
Additive decomposition of the
    total strain rate 38
Additive symmetry 19
Almansi strain tensor 14
Associated plastic flow 61

Baushinger effect 66
Biot strain tensor 23
Black body 29
Body forces 8
Boundary conditions
    of first kind 32
    of second kind 32
    of third kind 32
    of fourth kind 32
Brittle rupture theory 131
Bulk modulus 106

Capacity matrix 85
Cartesian coordinates 3, 8, 12, 13, 16
Cauchy–Green tensor 23
Cauchy relations 8
Cauchy strain 45
Cauchy stress 24
Compression test 170
Conductivity matrix 85
Conjugate variables 26, 38
Consistency 168
Convergence 107
Coordinates Cartesian 3, 12, 13, 16
    Euler 12
    Lagrange 12
Coordinates
    cylindrical 21
    material 12
    spatial 12
    spherical 7, 11, 32
Coupled thermo-plasticity 163
Crank–Nicolson scheme 86
Creep 117
    at constant uniaxial stress 117
    potential 124

Creep (*continued*)
    primary 117
    rupture 128
    secondary 117
    tertiary 117
    test 117
Creep theories 120, 124
    deformational type 124
    in complex stress state 124
Cryotherapy 94

Deformation gradient 13, 14
Deformation rate tensor 19, 20
Density 31
Deviatoric strain 9
    stress 5
Dissipation vector 165
Dissipative process 41
Divergence theorem 81
Drucker postulate 60
Ductile rupture theory 120

Effective strain 10
    stress 7
Einstein summation convention 3
Elasticity 48
Elastic limit 45
Elastic-plastic matrix 106
Elastic strain rate 48
Elasto-plasticity tensor 74
Elasto-visco-plasticity 78
Electroslag welding 97
Entropy 40
Equilibrium condition 24
Euclidean norm 169
Eulerian triads 13, 25
Euler scheme 86
Euler's theorem 82
Extension with torsion 17

Finite element solution 80
    of heat flow equations 80
    of Navier–Stokes equations 90
    of thermo-elasto-plastic problems 100

Finite strain 12
   models 151
   tensor 12
Flow theories 55
Fourier law 30
Friction coefficient 168
   rate sensitivity index 168

Gaussian quadrature points 84
Gauss–Green theorem 163
Gibbs equation 41
Green–Lagrange strain rate tensor 26
   strain tensor 14
Grey body 29

Hardening parameter 65
Heat conduction 27
   coefficient 28
Heat convection 28
   coefficient 28
Heat flow with phase change 94
Heat flux 27
   vector 30
Heat radiation 29
Heaviside function 93
Hencky–Iljuszyn deformation theory 53
Hencky strain 18, 23
Heredity theory 123
Hooke's law 73, 79
Huber–Mises yield condition 52

Ideal plasticity 52
Infinite series 18
Initial configuration 35
Initial temperature 32
Integrality equations 10, 11
Integration 104
Internal dissipation function 166
Internal variables 40
Irreversible process 35
Isotropic hardening 65, 73
Iterative accumulation 105

Jaumann stress 160

Kinematic hardening 66, 76
Kirchhoff stress 24, 26
Kronecker symbol 5
Kuhn–Tucker loading conditions 156

Lagrangian triads 13
Lame constants 48
Levy–Mises flow theory 55
Lie derivative 156
Linear thermal expansion tensor 49
Logarithmic strain tensor 18

Material
   Ideal elasto-plastic 50
   linear hardening elasto-plastic 50
   linear hardening rigid-plastic 50
   rigid ideal-plastic 50
   work hardening elasto-plastic 50
   work hardening plastic 50
   work hardening rigid-plastic 50
Mean stress 5, 53
Metric tensor 13, 15
Mid-point scheme 86
Multiplicative decomposition of the deformation gradient 38, 151, 155

Nadai–Davis theory 59
Navier–Stokes equation 33
Necking point 45
Newton equation 28
Non-isothermal plastic flow 153
Nonlinear heat conduction 84
Norm of residual equations 169

Odqvist hypothesis 65
One-step scheme 85
Orthogonal system 16
Orthogonal tensor 13

Phase change 94
Piola–Kirchhoff stress tensor
   first 24
   second 24
Plastic flow theory 55
Plastic modulus 76
Plastic potential function 52, 68
Plastic potential rate 73
Plasticity 50
Poisson ratio 48
Polar decomposition theorem 13
Prandtl–Reuss theory 55
Principal strains 9
   invariants of the strain tensor 9
   invariants of the deviatoric strain 10
Principal stresses 5
   invariants of the stress tensor 5
   invariants of the deviatoric stress 6
Principal stretches 13, 18
Process
   active 57, 63
   neutral 57, 63
   passive 57, 63
   simple 57
Proportional limit 45

Quinney–Taylor hypothesis 65

Radiation coefficient 29
Ramberg–Osgood elasto-plastic material 50

Reference temperature 29
Relaxation 117
  test 117
Remaining specific energy 41
Ring compression 170
Rotation matrix 160
Rupture 128

Saint-Venant de. integrality
    equations 10
Shear modulus 48
Shear stress 168
Skew-symmetric tensor 20
Small strain thermo-elasto-
    plasticity 71
Spatial description 36
Specific energy 40
Specific free enthalpy 40
Specific heat 31
  at constant volume 164
Specific reversible work 38
Spherical coordinates 7, 11, 32
Spherical strain 9
Spherical stress 5
Spin tensor 23
Stability analysis 88
  of plastic material 60
Stefan–Boltzmann law 29
Strain 8
Strain hardening theory 121
Strain tensor 8
  Almansi 14
  Cauchy 8
  Green 14
  infinitesimal 20
Stress 3
  and time functions 120
  functions 110
  measure 24
  tensor 3
  vector 3
Stress–strain curve 45
Stress–strain relations 48
Surface tractions 157, 159
Symmetric tensor 13, 20

Tangent stiffness matrix 106

Temperature 27
  functions 119
  gradient 30
Tensor
  infinitesimal strain 8
  logarithmic strain 18
  first Piola–Kirchhoff 24
  second Piola–Kirchhoff 24
Thermal diffusion coefficient 31
Thermal load vector 165
Thermal strain rate 48
Thermomechanical coupling
    matrix 167
Thermomechanical process 35
Thermo-rigid plastic 168
Thermo-rigid visco-plastic 168
Time hardening theory 121
Time functions 117
Time integration schemes 84
Total Lagrange formulation 157
Total strain theory 120
Translated stress deviator 76
Translated stress tensor 76
Trapezoidal scheme 86
Two-point tensor 13
Two-step schemes 87

Updated Lagrange formulation 158
Updated Lagrange–Hughes
    formulation 160
Updated Lagrange–Jaumann
    formulation 158

Variational formulation 82
Velocity of particle 19
Viscosity 33
  coefficient 79

Weighted residual method 80
Work-hardening 63, 65, 75

Yield criteria
  Huber–Mises 52
  Tresca–Guest 52
Yield point 45
Yield surface 52
  cylindrical 53
Young's modulus 48